Lecture Notes in Mathematics

A collection of informal reports and seminars
Edited by A. Dold, Heidelberg and B. Eckmann, Zürich

281

G. M. Kelly, M. Laplaza,
G. Lewis, and S. Mac Lane

Coherence in Categories

Edited by Saunders Mac Lane, University of Chicago, IL/USA

Springer-Verlag
Berlin · Heidelberg · New York 1972

AMS Subject Classifications (1970): 1802, 18A25, 18D15, 18D25

ISBN 3-540-05963-6 Springer-Verlag Berlin · Heidelberg · New York
ISBN 0-387-05963-6 Springer-Verlag New York · Heidelberg · Berlin

Offsetdruck: Julius Beltz, Hemsbach/Bergstr.

PREFACE

Categorical arguments are full of diagrams of arrows, and in very many cases these diagrams commute. Coherence theorems are theorems which state that a large class of diagrams always commute, or theorems which describe conditions sufficient to insure commutativity. The first such coherence result dealt with the case of a tensor product functor $\otimes$ which is associative, but only up to a natural isomorphism

$$a : A\otimes(B\otimes C) \cong (A\otimes B)\otimes C \quad ,$$

One of the diagrams involving this associativity a is the pentagon

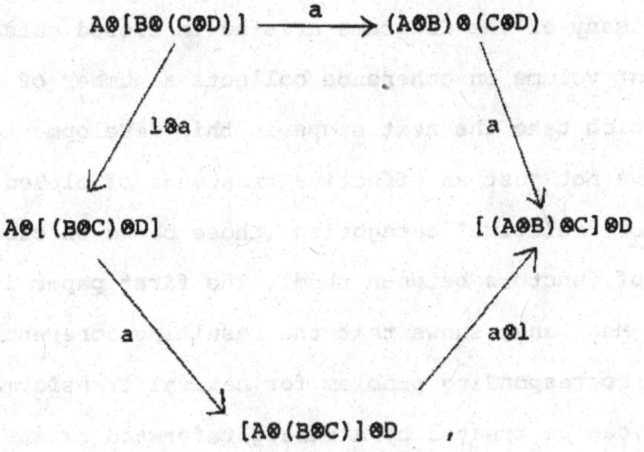

which requires that both possible reassociations of four factors are equal. The first coherence theorem asserts that if this pentagon commutes, then all (larger) diagrams involving this associativity will also commute. This result was first obtained by J.D. Stasheff in his studies of higher homotopies in algebraic topology (Trans. A.M.S. 108(1963), 275-282), and was also established independently by S. Mac Lane and by D.B.A. Epstein. Their two papers (and the others mentioned later in this preface) are those cited in the bibliography of the article by G. M. Kelly on <u>Many Variable Functorial Calculus</u>, in this volume.

The next major coherence question was that for diagrams

involving both a tensor product functor and an (internal) hom functor; categories of vector spaces or of modules over commutative rings present such diagrams. The general case is that of a closed category; the work of Eilenberg-Kelly brought out the great utility of their study but involved very many complicated commuting diagrams. This emphasized the importance of getting a good coherence theorem. Lambek obtained a preliminary coherence result which recognized the connection between these questions and the cut-elimination theorem of Gentzen-style proof theory, and then Kelly-Mac Lane combined some of Lambek's ideas with the notion of the "graph" of a generalized natural transformation to obtain a much more extensive coherence theorem covering many of the diagrams arising in closed categories.

The present volume on coherence collects a number of closely related papers which take the next steps in this development. First, one wishes to have not just an effective treatment of closed categories but one for "relative" categories (those based on closed categories) and of functors between them. The first paper in this volume, by Kelly-Mac Lane, shows that the resulting coherence problem, and the corresponding problem for natural transformations between functors,can be treated by a straight forward extension of the Kelly-Mac Lane closed category method. In similar spirit Geoffrey Lewis treats a coherence for a "closed functor" between two closed categories; here interesting new phenomena arise because it is by no means the case that "all" diagrams commute; the problem is rather that of giving necessary and sufficient conditions for commutativity. Finally, Kelly,in his study here of the cut-elimination theorem, shows that this classical result really provides for the preservation of coherence under the addition of adjoints; the original case being the coherence of tensor products and the addition to ⊗ of a right adjoint, the internal hom, one for each functor - ⊗ B. This general understanding is formulated against the background concepts provided by Kelly's paper "An Abstract Approach to Coherence"; there it is indicated how each coherence problem can be described in terms of the "actions" of a suitable structural category called a "club" (Each club is a monoid in a comma category). Moreover, this development in turn requires a generalized functor category, where the objects are not only the usual functors $A \longrightarrow B$ of one variable, but also functors $A\times\cdots\times A \longrightarrow B$ of n variables, for all n. This concept, which is of independent interest, is set forth in Kelly's first paper in this volume "Many Variable Functorial Calculus I"

Monoidal functors provide another fascinating coherence problem, treated by Lewis in his paper in this volume "Coherence for a Closed Functor". A monoidal functor $\phi: V \longrightarrow V'$ between two categories V and V', each with an associative tensor product, is a functor between the underlying categories together with a natural transformation

$$\tilde{\phi}: \phi A\otimes\phi B \longrightarrow \phi(A\otimes B) \quad .$$

This transformation need not be an isomorphism, as for instance in the case of the forgetful functor $U: K\text{-}\underset{\sim}{Mod} \longrightarrow \underset{\sim}{Ab}$, (K a commutative ring) where there is a map

$$\tilde{\phi}: UA\otimes UB \longrightarrow U(A\otimes_K B)$$

not necessarily an isomorphism. One assumes, with the given associativities a and a' in V and V', that the diagram

$$\begin{array}{ccccc} \phi A\otimes(\phi B\otimes\phi C) & \xrightarrow{1\otimes\tilde{\phi}} & \phi A\otimes\phi(B\otimes C) & \xrightarrow{\tilde{\phi}} & \phi(A\otimes(B\otimes C)) \\ \downarrow a' & & & & \downarrow \phi a \\ (\phi A\otimes\phi B)\otimes\phi C & \xrightarrow{\tilde{\phi}\otimes 1} & \phi(A\otimes B)\otimes\phi C & \xrightarrow{\tilde{\phi}} & \phi((A\otimes B)\otimes C) \end{array}$$

commutes. The coherence question is: Do all bigger such diagrams commute? Moreover, does the same hold when we add identities I and I' for each $\otimes$ product with natural isomorphisms $b: I\otimes A \cong A$, $b': I'\otimes A' \cong A'$, $\gamma: A\otimes I \cong A$, and when we add to the data $(\phi, \tilde{\phi})$ for the monoidal functor above a morphism $\phi^\circ: I' \longrightarrow \phi I$ such that

$$\begin{array}{ccc} I'\otimes\emptyset A & \xrightarrow{b'} & \emptyset A \\ {\scriptstyle \emptyset^\circ\otimes 1}\downarrow & & \uparrow{\scriptstyle \emptyset b} \\ \emptyset I\otimes\emptyset A & \xrightarrow{\tilde{\emptyset}} & \emptyset(I\otimes A) \end{array} \qquad\qquad \begin{array}{ccc} \emptyset A\otimes I' & \xrightarrow{r'} & \emptyset A \\ {\scriptstyle 1\otimes\emptyset^\circ}\downarrow & & \uparrow{\scriptstyle \emptyset r} \\ \emptyset A\otimes\emptyset I & \xrightarrow{\tilde{\emptyset}} & \emptyset(A\otimes I) \end{array}$$

both commute. For the answers to these questions and the corresponding questions for closed categories, we refer to the paper by Lewis.

Distributivity involves a different sort of coherence question, which has been raised by Hyman Bass (in conversations) and which has been studied by J. Benabou (results not yet published). Given are two functors, direct sum $\oplus$ and tensor product $\otimes$, with a natural homomorphism

$$d : A\otimes(B\oplus C) \longrightarrow (A\otimes B)\oplus(A\otimes C) \quad .$$

One again asks for a certain number of basic identities which will suffice to get the coherence of all diagrams involving this distributivity d -- and involving as well the associativity isomorphism for both $\oplus$ and $\otimes$. The first paper by Laplaza in this volume tackles this problem. Here again, it is not that all diagrams commute; Laplaza's second paper in this volume, building on his first, gives a definitive necessary and sufficient condition for commutativity.

There are many further problems, in particular those noted by Kelly in his papers here. We trust that this volume may focus interest and encourage further progress.

Saunders Mac Lane

June 6, 1972

TABLE OF CONTENTS

CLOSED COHERENCE FOR A NATURAL TRANSFORMATION *

G. M. KELLY
The University of New South Wales, Australia
and
SAUNDERS MAC LANE
The University of Chicago, Chicago, Ill., USA
Received February 23, 1972

1. INTRODUCTION

For two functors F, $G:\underline{A} \longrightarrow \underline{A}'$, the definition of a natural transformation $k:F \longrightarrow G$ states that k assigns to each object A of $\underline{A}$ an arrow $k_A:FA \longrightarrow GA$ such that the diagram

$$\begin{array}{ccc} FA & \xrightarrow{k_A} & GA \\ {\scriptstyle Ff}\downarrow & & \downarrow{\scriptstyle Gf} \\ FB & \xrightarrow[k_B]{} & GB \end{array}$$

commutes for every morphism $f:A \longrightarrow B$ of $\underline{A}$. This one condition implies that "all" other appropriate diagrams commute; more explicitly, all the diagrams built up from k, F, G, and a string $A \longrightarrow B \longrightarrow C \longrightarrow \cdots \longrightarrow E$ of composable morphisms of $\underline{A}$. This fact, stated with suitable precaution, is a coherence theorem for a natural transformation. This paper is concerned with proving such a coherence theorem when $\underline{A}$ and $\underline{A}'$ are not ordinary categories, but categories based on some closed category $\underline{V}$ (and when F and G are correspondingly $\underline{V}$-functors).

The resulting theorem will include our recent coherence theorem [1] for a closed category. That theorem has made unnecessary the many

* This paper is based on research supported by grants from the National Science Foundation and the Louis Block Fund of the University of Chicago.

individual verifications as in [3] of the commutativity of large diagrams needed to establish the fundamental properties of closed categories. Our present result similarly avoids the even more horrendous diagrams involving not only the closed category $\underline{V}$ but also extra data such as $\underline{V}$-categories $\underline{A}$ and $\underline{A}'$, $\underline{V}$-functors $F,G:\underline{A} \longrightarrow A'$ and a $\underline{V}$-natural transformation $k:F \longrightarrow G$. The surprising fact is that the methods of our previous coherence proof [1] apply exactly to the present case, so that our proof will also show how similar coherence theorems could be established for more complex data over a closed category -- say for several $\underline{V}$-functors with various natural transformations between them.

Our purpose requires us to use only the basic definitions of a $\underline{V}$-category, a $\underline{V}$-functor etc., and not their properties as established in [3]; for the proofs of these latter involve just such verifications of commutativity as we aim to make redundant. So we "do not know" that a $\underline{V}$-category $\underline{A}$ has an underlying ordinary category $\underline{A}_0$, with a $\underline{V}$-valued hom-functor $\underline{A}_0{}^{op} \times \underline{A}_0 \longrightarrow \underline{V}$; still less that there is a $\underline{V}$-category $\underline{A}^{op} \otimes \underline{A}$ and a $\underline{V}$-functor $Hom:\underline{A}^{op} \otimes \underline{A} \longrightarrow \underline{V}$. For this reason, we replace the definition of $\underline{V}$-natural transformation given in [3], which used $\underline{A}_0$ and the functor $\underline{A}_0{}^{op} \times \underline{A}_0 \longrightarrow \underline{V}$, by a more basic one that is easily seen to be equivalent.

We consider first such a $k:F \longrightarrow G:\underline{A} \longrightarrow \underline{A}'$ when $\underline{V}$ is merely monoidal. The coherence in this case is proved by a purely combinatorial argument, like the now classical proof ([6], [7]) that commutativity of a certain pentagonal diagram implies coherence for associativity.

We then pass to the case of a closed $\underline{V}$, by adding the right adjoint $[X,-]$ of $-\otimes X$. The difficulty in the way of a similar proof is that now, when we form a composite $A \longrightarrow B \longrightarrow C$ of the morphisms that make up our diagrams, B may be arbitrarily much more complicated than A and C, so that the familiar kind of proof by induction is precluded. We find our salvation in a cut-elimination argument, directly generalizing that in our earlier paper, and inherited from Gentzen [4] via Lambek [5]. This argument asserts that the morphisms which constitute our diagrams can be built up from certain basic ones <u>without</u> recourse to the above operation of composition (Gentzen's "cut"). For the case of a closed category, there are four basic types: A <u>central</u> morphism (a combination of associativities and commutativities), a <u>tensor product</u> of two morphisms already constructed, an <u>adjoint</u> $A \longrightarrow [B,C]$ of a given morphism $A \otimes B \longrightarrow C$, or a process of applying <u>evaluation</u> e after a given morphism $f:A \longrightarrow B$ as in

$$[B,C] \otimes A \xrightarrow{1\otimes f} [B,C] \otimes B \xrightarrow{e} C.$$

(Actually we need a more complex process of evaluation which acts on two given morphisms $f:A \longrightarrow B$ and $g:C \otimes D \longrightarrow E$ to form

$$[B,C] \otimes A \otimes D \xrightarrow{1\otimes f\otimes 1} [B,C] \otimes B \otimes D \xrightarrow{e} C \otimes D \xrightarrow{g} E.)$$

These four types are used as follows: There is a prospective morphism

$$[B,C] \otimes A \otimes C \otimes [R,S] \longrightarrow [B,A] \otimes S \otimes \cdots$$

with given domain and codomain. These data suggest that such a morphism might in fact be constructed by one or more of the four basic types of construction. The problem is thereby reduced to the construction of simpler prospective morphisms. The basic idea of the proof in [1] is to show that this construction process, if possible, does yield a unique result, provided the proposed domain and codomain, regarded as shapes (=formulas) are "proper"; that is, that they never involve an internal-hom of the form [nonconstant, constant]. Our present argument is the same process, with just one more type of construction, corresponding essentially to the composition of a string, as described above.

This development may cast some light on the surprising use in category theory of the proof-theoretic method of cut elimination.

We recall that Lambek in [1] started this use; our work has

also been stimulated by knowledge of the results of M.E. Szabo (not yet published) and of an early form of the results of G. Lewis (now published in this volume).

2. STATEMENT OF RESULTS

Let $\underline{V}$ be a monoidal category, with the usual tensor product $\otimes$ and the usual natural isomorphisms a (associativity), c (commutativity) and $b = b_A : A \otimes I \cong A$, as specified for example in [1]. A $\underline{V}$-category $\underline{A}$ is a set of objects A, B,$\cdots$, a function which assigns to each pair of objects A, B a "hom-object" (A,B) or $\underline{A}(A,B)$ which is an object of $\underline{V}$, together with two families of morphisms

$$M = M^B_{AC} : (B,C) \otimes (A,B) \longrightarrow (A,C), \qquad j = j_A : I \longrightarrow (A,A)$$

such that the familiar diagrams

$$\begin{array}{ccccc} \{(C,D)\otimes(B,C)\}\otimes(A,B) & \xrightarrow{a} & (C,D)\otimes\{(B,C)\otimes(A,B)\} & \xrightarrow{1\otimes M} & (C,D)\otimes(A,C) \\ \downarrow{\scriptstyle M\otimes 1} & & & & \downarrow{\scriptstyle M} \\ (B,D)\otimes(A,B) & & \xrightarrow{\quad M \quad} & & (A,D), \end{array} \qquad (2.1)$$

$$\begin{array}{ccc} (A,B)\otimes I & \xrightarrow{1\otimes j_A} & (A,B)\otimes(A,A) \\ & \searrow{\scriptstyle b} & \downarrow{\scriptstyle M} \\ & & (A,B), \end{array} \qquad \begin{array}{ccc} I\otimes(A,B) & \xrightarrow{j_B\otimes 1} & (B,B)\otimes(A,B) \\ & \searrow{\scriptstyle bc} & \downarrow{\scriptstyle M} \\ & & (A,B) \end{array} \qquad (2.2)$$

always commute. Let $\underline{A}'$ be a second such category, with hom-objects written (A',B')' or (A',B'). A $\underline{V}$-functor $F:\underline{A} \longrightarrow \underline{A}'$ is then a pair of functions which assign to each object A of $\underline{A}$ an object FA of $\underline{A}'$ and to each pair of objects A, B of $\underline{A}$ a morphism $F_{A,B} : (A,B) \longrightarrow (FA,FB)$ of $\underline{V}$ such that the familiar diagrams

$$\begin{array}{ccc} (B,C)\otimes(A,B) & \xrightarrow{M} & (A,C) \\ \downarrow{\scriptstyle F\otimes F} & & \downarrow{\scriptstyle F} \\ (FB,FC)\otimes(FA,FB) & \xrightarrow{M'} & (FA,FC), \end{array} \qquad \begin{array}{ccc} I & \xrightarrow{j_A} & (A,A) \\ & \searrow{\scriptstyle j'_{FA}} & \downarrow{\scriptstyle F} \\ & & (FA,FA) \end{array} \qquad (2.3)$$

always commute. These definitions apply also when $\underline{V}$ is a closed category (add [-,-]).

If F and $G:\underline{A}\longrightarrow\underline{A}'$ are two $\underline{V}$-functors, a $\underline{V}$-natural transformation $k:F\longrightarrow G:\underline{A}\longrightarrow\underline{A}'$ is a function k which assigns to each object A of $\underline{A}$ a morphism $k_A:I\longrightarrow(FA,GA)$ of $\underline{V}$ such that every diagram

$$\begin{array}{ccc} (A,B) & \xrightarrow{\;F\;} & (FA,FB) \\ {\scriptstyle G}\downarrow & & \downarrow{\scriptstyle k^\flat_{FA,B}} \\ (GA,GB) & \xrightarrow[\;k^*_{A,GB}\;]{} & (FA,GB) \end{array} \qquad (2.4)$$

commutes, where $k^\flat = k^\flat_{A',B}$ and $k^* = k_{A,B'}$ are defined from k by

$$\begin{array}{ccc} (A',FB) & \xrightarrow{\;k^\flat\;} & (A',GB) \\ {\scriptstyle bc}\uparrow & \cdot & \uparrow{\scriptstyle M'} \\ I\otimes(A',FB) & \xrightarrow[\;k_B\otimes 1\;]{} & (FB,GB)\otimes(A',FB), \end{array} \qquad (2.5)$$

$$\begin{array}{ccc} (GA,B') & \xrightarrow{\;k^*\;} & (FA,B') \\ \uparrow{\scriptstyle b} & & \uparrow{\scriptstyle M'} \\ (GA,B')\otimes I & \xrightarrow[\;1\otimes k_A\;]{} & (GA,B')\otimes(FA,GA). \end{array} \qquad (2.6)$$

When $\underline{V}=\underline{Set}$ with $\otimes = \times$, then I is a one-point set and we may regard k_A as a morphism $FA\longrightarrow GA$ in $\underline{A}'$. Then $k^\flat_{A',B}$ is just composition with k_B on the left, $k^*_{A,B'}$ is composition with k_A on the right, and the first commutative diagram above is just the familiar definition of naturality as given in our introduction -- or as in [3], page 466.

If the diagonal morphism of the commutative diagram (2.4) is called $v_{A,B}:(A,B)\longrightarrow(FA,GB)$, then a natural transformation can be defined to be a family of such morphisms v_{AB} such that the diagrams

$$M'\circ(v\otimes F) = v\circ M = M'\circ(G\otimes v):(B,C)\otimes(A,B)\longrightarrow(FA,GC) \qquad (2.7)$$

all commute. Then k may be constructed from v as $k_A = v_{A,A}\circ j_A:I\longrightarrow(FA,GA)$. An easy diagram chase shows the equivalence of these two definitions of naturality. The definition by k is more convenient for our purposes.

The whole situation $W = \{\underline{V}, \underline{A}, \underline{A}', F, G, k\}$ consisting of the $\underline{V}$-natural transformation k and the accompanying data will be called a monoidal naturality. If in addition the category $\underline{V}$ is closed, in the usual sense that $-\otimes X:\underline{V} \longrightarrow \underline{V}$ has a right adjoint $[X,-]$, with unit $d:X \longrightarrow [Y,X\otimes Y]$ and counit $e:[X,Y] \otimes X \longrightarrow Y$ for the adjunction, we shall call the whole situation a closed naturality. We abbreviate both terms to naturality.

We can now define a certain typical or "formal" naturality, in terms of "shapes". These shapes, used in [1], were the well-formed formulas for the functors obtained by iterated composition from $\otimes$ and $[\ ,\ ]$; thus the shape I is the formula for the functor constantly I, and 1 is the formula for the identity $\underline{V} \longrightarrow \underline{V}$. In general, the shapes S and T were defined recursively by requiring that I, 1, $T \otimes S$ and $[T,S]$ are shapes. We retain this notion, and introduce also a new one, the N-shapes, to include functors built up using also (A,B), (A',B'), (FA,B'), (GA,GB), etc. We will write - and -' instead of variables A and B'. Formally, we then define our N-shapes by adding to the previous four inductive rules S1 to S4 for shapes one new such rule:

S5 (-,-) and (x,y)' are N-shapes, where each of x and y is any one of -', F-, or G- (nine cases).

These N-shapes S5 we call the terms. Each N-shape is to have its set of N-variables, where an N-variable-set is now to be a finite ordinal number $n = \{1, 2, \cdots, n\}$ equipped with two functions

$$\text{Variance: } n \longrightarrow \{\text{co, contra}\}, \text{ Type: } n \longrightarrow \{\underline{V}, \underline{A}, \underline{A}'\}.$$

The ordinal sum n + m and the twisted sum $n \mathbin{\tilde{+}} m$ (change the variances of all variables in n) are defined as before. With each N-shape T is associated an N-variable set v(T) defined just as before except that the previous rule V2 for the shape 1 is replaced by the (evidently intended) rules

V2: v(1) is the ordinal 1, covariant, type $\underline{V}$

V5: v(-,-) is the ordinal 2, contra, co, both of type $\underline{A}$,

v(x,y)' is the ordinal 2, contra, and co, where x = -' has type $\underline{A}'$, x = F- has type $\underline{A}$ and x = G- has type $\underline{A}$.

For N-shapes T and S an N-graph $\xi : T \longrightarrow S$ is to be a fixed-point-free involution on the disjoint union v(T) + v(S), with the property that mates under the involution ξ have the same type and opposite variance in the twisted sum $v(T) \mathbin{\tilde{+}} v(S)$. The composition of N-graphs, and compatible N-graphs, are defined as before. With the evident descriptions of a tensor product $\xi \otimes \eta$ and a bracket $[\xi,\eta]$ of two N-graphs ξ, η we obtain a closed category $\underline{V}_0$ of N-graphs, with objects all N-shapes and morphisms all N-graphs (This category $\underline{V}_0$ is the N-analog of the category called $\underline{G}$ in [1]). Moreover, the integral N-shapes (those formed without using square brackets) form in the same way a monoidal category $\underline{V}_{00}$.

Over $\underline{V}_0$ we then obtain a "typical" closed naturality, as follows. First, construct a $\underline{V}_0$-category $\underline{A}_0$ with one object -, with hom object the shape (-,-), and with composition and unit morphisms the evident N-graphs

$$(y,z) \otimes (x,y) \longrightarrow (x,z), \qquad I \longrightarrow (x,x), \tag{2.8}$$

where we have replaced the blanks in the shapes by letters, with mates under the intended graph represented, in the familiar way, by the same letters. Take $\underline{A}_0'$ to be the $\underline{V}_0$-category with three objects, written -', F-, and G-, with the corresponding shapes (the nine terms (x,y) introduced in S5 above) as hom-objects in $\underline{V}_0$, and with the definition of composition M' and unit j' like those just given for $\underline{A}_0$. These descriptions yield $\underline{V}_0$-categories $\underline{A}_0$ and $\underline{A}_0'$, and there are $\underline{V}_0$-functors F_0 and G_0; for example, F_0 sends the sole object written - in $\underline{A}_0$ to the object denoted F- in $\underline{A}_0'$, and the arrow F_{0--} is the graph $(x,y) \longmapsto (Fx,Fy)$. Finally, the graph $I \longrightarrow (Fx,Gx)$ from the shape I to the shape (F-,G-) is $\underline{V}_0$-natural. Thus we have defined a closed naturality W_0, consisting of $\underline{V}_0$, $\underline{A}_0$, $\underline{A}'_0$, and $k_0:F_0 \longrightarrow G_0$. Upon restriction to integral N-shapes, we similarly obtain a monoidal naturality W_{00}.

The closed category $\underline{V}_0$ in the naturality W_0 has N-shapes as objects and N-graphs as morphisms; it is the exact analog of the category $\underline{G}$ of graphs as used in [1]. The diagrams which commute will lie in a different closed category $\underline{V}_\#$, which depends on $\underline{V}$ and on the naturality W, and will be the analog of $\underline{N}(\underline{V})$ in [1]. The objects of that category $\underline{N}(\underline{V})$ were still the shapes; this was just the usual precaution needed to state any coherence theorem: The diagrams which commute have vertices not objects of $\underline{V}$ but formulas (shapes) for objects of $\underline{V}$. The morphisms $S \longrightarrow T$ of the category $\underline{N}(\underline{V})$ were the natural transformations $|S| \longrightarrow |T|$ between the associated functors. This depended on the observation that each (ordinary) shape T in n variables defines for each closed category $\underline{V}$ a functor $|T| = |T|_V$ of n variables (in $\underline{V}$). We will make the corresponding constructions of functors in the present case; the natural transformations between these will provide the morphisms for $\underline{V}_\#$, over which we then get a naturality $W_\#$.

This program meets an obstacle because, while $\underline{V}$ is a category, $\underline{A}$ and $\underline{A}'$ are not _a priori_ given as categories at all, so that we cannot interpret, say, the shape (-,-) as a functor. We do not want to assume the fact that each $\underline{V}$-category $\underline{A}$ has an underlying (ordinary) category $\underline{A}_0$, because our coherence result, suitably formulated, can be used to simplify the proof of this fact. We eliminate the obstacle by regarding $\underline{A}$ and $\underline{A}'$ henceforth as the _discrete_ categories determined by their objects. The reason we can do so without loss is that our coherence result is only concerned with diagrams in $\underline{V}$ and makes no reference to morphisms in $\underline{A}$ or in $\underline{A}'$.

Let W be a naturality with $\underline{A}$ (and hence $\underline{V}$ and $\underline{A}'$) non-empty.

Let T be an N-shape with variable-set n. For each i in n take B_i equal to the category (type i) or the category (type i)^{op} according to the variance of i in n. Now use T to define a functor

$$|T| = |T|_W : \underline{B}_1 \times \cdots \times \underline{B}_n \longrightarrow \underline{V}$$

by recursion in the evident way suggested by the notation for the shape T. We can in fact take over our previous recursive definition of |T| ([1], page 103 foot) upon adding the appropriate rules for our new terms. Thus the shape (-,-) leads to the functor $|(-,-)| : \underline{A}^{op} \times \underline{A} \longrightarrow \underline{V}$ given by $A,B \longmapsto (A,B)$; this is trivially functorial because $\underline{A}$ is discrete. Similarly |(-',-')| is the corresponding hom-object functor for $\underline{A}'$, and |(F-,G-)| is the composite functor

$$\underline{A}^{op} \times \underline{A} \xrightarrow[F^{op} \times G]{} \underline{A}'^{op} \times \underline{A}' \xrightarrow[|(-,-)|]{} \underline{V};$$

here G is the object-map $\underline{A} \longrightarrow \underline{A}'$ considered as a functor. Similar definitions apply to the other seven terms of S5 above.

Given two N-shapes T and S, a natural transformation $f : |T| \longrightarrow |S|$ now consists of an N-graph $\Gamma f = \xi : T \longrightarrow S$ and certain components, as in [1]. These components are to satisfy the usual naturality conditions, which are vacuous for variables from the (discrete) categories $\underline{A}$ and $\underline{A}'$, but not for those from $\underline{V}$. The composite gf of two natural transformations is again defined as in [1], with one precaution. When the graphs η and ξ of g and f are not compatible, there are certain variables which occur in closed loops (as described in [1]). To get a definite composite, one must choose a value for each such variable X_i; in [1] we took $X_i = I$, the base object of $\underline{V}$. Now we also have "variables" in $\underline{A}$ and $\underline{A}'$; we select a fixed object D in the non-empty category $\underline{A}$, and we define a composite by specializing the variables in $\underline{V}$ to I, those in $\underline{A}$ to D, and those in $\underline{A}'$ to FD. The resulting composites of natural transformations will then depend on the choice of D, but for fixed D the composition is still associative and determines a closed category $\underline{V}_{\#}$ with objects the N-shapes T, S,$\cdots$ and morphisms $T \longrightarrow S$ all the natural transformations $|T|_W \longrightarrow |S|_W$. It will turn out that the dependence on D does not matter (we will actually use only the "compatible" part of $\underline{V}_{\#}$).

Next we construct a closed naturality $W_{\#} = \left\{ \underline{V}_{\#};\ F_{\#},\ G_{\#}:\underline{A}_{\#} \longrightarrow \underline{A}'_{\#};\ k_{\#}:F_{\#} \longrightarrow G_{\#} \right\}$. Here the $\underline{V}_{\#}$-category $\underline{A}_{\#}$ is to have one object, the "variable" -, with hom-object the shape (-,-) regarded as an object of $\underline{V}_{\#}$; composition is the natural transformation with the graph already described in (2.8) but with components given in the evident way by the actual composition operation M in the given $\underline{V}$-category $\underline{A}$. The unit map j for $\underline{A}_{\#}$ is constructed in a similar way. Corresponding constructions yield a $\underline{V}_{\#}$-category $\underline{A}'_{\#}$ with three objects -', F-, and G-. Then $F_{\#}:\underline{A}_{\#} \longrightarrow \underline{A}'_{\#}$ is the $\underline{V}_{\#}$-functor which is $(-) \longmapsto F-$ on objects, while $\underline{A}_{\#}(-,-) \longrightarrow \underline{A}'_{\#}(F-,F-)$ is that morphism in $\underline{V}_{\#}$ which has the graph F_{0--} already described and the components

$$F_{A,B}:\underline{A}(A,B) \longrightarrow \underline{A}'(FA,FB),$$

for each pair of objects A, B of $\underline{A}$. Thus $F_{\#}$, and similarly $G_{\#}$, is a sort of transliteration of F to act on N-shapes; it is routine to verify that it is a $\underline{V}_{\#}$-functor. Finally, $k_{\#}:F_{\#} \longrightarrow G_{\#}$ is the natural transformation with graph $I \longrightarrow (Fx,Gx)$ and with components $I \longrightarrow \underline{A}(FA,GA)$ the components k_A. Thus we have defined a closed naturality $W_{\#}$; moreover, the process "take the graph" defines a morphism $\Gamma:W_{\#} \longrightarrow W_0$ of naturalities which is strict (i.e., one that preserves all the naturality data on the nose). Again the same applies for monoidal shapes. We have in effect reproduced much of the structure of W within $W_{\#}$ (though $W_{\#}$ is much "smaller"); the diagrams which commute are to be those in $W_{\#}$ -- i.e., certain diagrams of arrows in $\underline{V}_{\#}$.

Given any naturality W, involving a closed category $\underline{V}$, the class of <u>central</u> morphisms of $\underline{V}$ is the smallest class of morphisms containing all identities for objects of $\underline{V}$, all instances of a, a^{-1}, b, b^{-1}, c, and closed under $\otimes$ and composition. The allowable morphisms of $\underline{V}$ are defined in the same way, but requiring also closure under [,] and including all instances of the unit d and counit e of the basic adjunction; i.e., all

$$d: X \longrightarrow [Y, X \otimes Y], \qquad e: [Y, X] \otimes Y \longrightarrow X$$

for objects X and Y of $\underline{V}$. These closure conditions are exactly the five conditions AM1 - AM5 of [1], except that in [1] they were applied only to a category in which the objects are shapes. We now introduce also the N-central morphisms, by adding to the closure conditions for central morphisms all instances of M, j, M', j', F, G, k, k* and $k^\flat$ (e.g., for every pair of objects A, B of $\underline{A}$, $F_{A,B}: (A,B) \longrightarrow (FA,FB)$ is N-central). We similarly define N-allowable by adding these same conditions to the definition of allowables. Observe that we can then construct a closed category $\text{Allow}_N(\underline{V})$ with objects all the objects of $\underline{V}$ and with only these morphisms which are N-allowable. The given naturality then restricts to a naturality (with the same $\underline{A}$, $\underline{A}'$, k) based not on $\underline{V}$ but on this closed subcategory $\text{Allow}_N(\underline{V})$. Briefly, the N-allowable morphisms of $\underline{V}$ are those, and only those, necessary for the description of the associativities, functors, natural transformations, etc. at issue.

In particular, the N-allowable morphisms of $\underline{V}_0$ are called the N-allowable N-graphs, those of $\underline{V}_\#$ the N-allowable natural transformations. With this language, we will prove that all the main theorems 2.1 - 2.4 of [1] hold in the present case (replacing "graph" by "N-graph" and "allowable" by "N-allowable"). In particular, we define a proper N-shape by the previous four conditions PS1 - PS4 (the essential restrictive condition is that [T,S] with T not constant and S constant is improper), adding the condition (for x,y = -', F- or G-):

PS5. Every term (-,-) or (x,y)' is proper.

The principal coherence theorem then states

Theorem 2.4_N. If W is any naturality, while T and S are proper N-shapes then any two N-allowable natural transformations $f, f' : |T|_W \longrightarrow |S|_W$ with the same N-graph are equal: $f = f'$.

In other words, any diagram of N-allowable maps between proper shapes commutes when the diagram of graphs commutes.

It will be clear from the proof that there is nothing to change if, instead of $k:F \longrightarrow G$, we are given only $F:\underline{A} \longrightarrow \underline{A}'$ or only $\underline{A}$, or only $\underline{A}$ and $\underline{A}'$.

3. THE CENTRAL MORPHISMS

The proofs of our coherence theorems will be constructed simply by extending the proofs of [1] to take care of the new case of those shapes which are terms. Indeed our argument will follow exactly the pattern and numbering of [1] with a minor technical variation: We consider the central morphisms (§4 there) before the "monoidal case".

We now define: An N-shape is <u>constant</u> when its construction involves no terms and no shapes 1; it is <u>integral</u> when no [,] are involved; it is <u>prime</u> when it is a term, a shape 1, or an N-shape of the form [T,S].

It is plausible to express any N-shape T as an iterated tensor product of its prime factors; we call this the <u>prime factorization</u> of T. Specifically, an easy recursion proves

PROPOSITION 3.1. There is a unique way of assigning to each N-shape T an integer $n \geq 0$ and a list $X_1, \cdots, X_n$ of prime N-shapes (the <u>prime factors</u> of T) in such a way that for each T exactly one of the following holds :

1° $n = 0$ and T is a constant integral N-shape,

2° $n = 1$ and $T = X_1$ is prime,

3° $n \geq 1$, $T = R \otimes S$, R has prime factors $Y_1, \cdots, Y_k$, S has prime factors $Z_1, \cdots, Z_m$, $n = k + m$ and

$$\left\{X_1,\cdots,X_n\right\} = \left\{Y_1,\cdots,Y_k,Z_1,\cdots,Z_m\right\}.$$

This prime factorization is exactly like that of [1], but is differently (and more simply) expressed; there we used the process of "substituting" prime shapes in an integral shape. Observe that in the shape $(-,-) \otimes I \otimes 1 \otimes [I,I \otimes I]$ there are three prime factors, including the constant prime factor $[I, I \otimes I]$, but that (the first) I here does not count as a prime factor.

Now let $\underline{K}$ stand for one of the categories $\underline{V}_0$ or $\underline{V}_\#$ and let $\Gamma:\underline{K} \longrightarrow \underline{V}_0$ be the functor which is the identity (when $\underline{K} = \underline{V}$) or the previous functor "take the graph" (when $\underline{K} = \underline{V}_\#$). In either case, the objects of $\underline{K}$ are N-shapes, and a central morphism can be described in the following way as a permutation of the prime factors.

PROPOSITION 3.2. Let $x:T \longrightarrow S$ be a central morphism in $\underline{K}$. Then T and S have the same number n of prime factors. If $X_1,\cdots,X_n$ are the prime factors of T and $Y_1,\cdots,Y_n$ those of S, then there is a permutation ξ of $\left\{1,\cdots,n\right\}$ such that $X_i = Y_{\xi i}$ and such that each variable of X_i has as mate under the graph of x exactly the same variable in $Y_{\xi i}$. If ξ' is another permutation of $\left\{1,\cdots,n\right\}$ which satisfies the same condition, then there is a permutation λ of $\left\{1,\cdots,n\right\}$ such that $\xi' = \lambda\xi$; moreover, $\lambda i \neq i$ implies that Y_i and $Y_{\lambda i}$ are equal constant shapes.

PROOF. The set of all those morphisms x of $\underline{K}$ for which this condition holds satisfies the three closure conditions AM1, AM3, and AM5 used in [1] to define central morphisms and therefore contains all central morphisms of $\underline{K}$. As for the uniqueness of ξ or λ, the graph of x will determine ξ except for those prime factors X_i which contain no variables -- i.e., the possible constant prime factors such as $[I,I]$, $[I \otimes I, I]$, etc. Thus we get the asserted permutation λ.

This proposition serves to replace Propositions 4.3 and 4.6 of [1]; it has the same corollaries (4.4 and 4.5). We will write $T \cong S$

for a (usually unnamed) central morphism of $\underline{K}$, and we then call the N-shapes T and S <u>centrally congruent</u>. Also two morphisms f, f' in $\underline{K}$ will be called <u>centrally congruent</u> (in symbols, $f \cong f'$) when there are central morphisms x and y such that f' = xfy. Any shape T with the n prime factors $X_1,\cdots,X_n$ is centrally congruent to

$$T \cong X_1 \otimes \cdots \otimes X_n \qquad (\text{or, } T \cong I \text{ if } n = 0) \qquad (3.1)$$

where the parentheses in this iterated tensor product all start (say) at the left; for that matter, there is a central congruence (3.1) which does not involve the commutativity isomorphism c.

The familiar coherence theorem [6] for a (symmetric) monoidal category translates at once to the present situation, replacing monoidal "variables" by arbitrary prime shapes, provided we observe, as in Lemma 4.7 of [1], that each constant prime is isomorphic to I. The result is

PROPOSITION 3.3. Let W be a closed naturality, while T and S are N-shapes with the prime factors $X_1,\cdots,X_n$ and $Y_1,\cdots,Y_n$, respectively, while ξ is a permutation of $\{1,\cdots,n\}$ such that $Y_{\xi i} = X_i$. Then there is a unique central morphism $f:T \longrightarrow S$ in $\underline{K}$ for which the graph mates each variable of X_i with the same variable in $Y_{\xi i}$.

This applies for $\underline{K} = \underline{V}_0$ or $\underline{V}_\#$, hence implies

THEOREM 3.4. If W is a closed naturality, each central N-graph $\varphi:T \longrightarrow S$ in $\underline{V}_0$ is the graph of a unique central natural transformation $f:T \longrightarrow S$ in $\underline{V}_\#$.

For any such f we call $Y_{\xi i}$ the prime factor of S <u>associated</u>, via f, with the prime factor X_i of T. With this notion we have the analogs of Propositions 4.10 and 4.11 of [1] (with slightly different proofs).

4. THE MONOIDAL CASE

We come to the combinatorial part of our coherence theorem. In the ordinary monoidal case (Theorem 3.1 of [1]), every graph $\xi:T \longrightarrow S$ between (integral) shapes was central, and there was one and only one natural transformation $f:T \longrightarrow S$ of graph ξ. The first of these properties fails for integral N-shapes, while the second holds only for N-central graphs. For example, the N-graphs (mates indicated by the same letters)

$$(x,y) \longrightarrow (fx,Fy),\quad (x,y) \longrightarrow (Gx,Gy),\quad (x,y) \longrightarrow (Fx,Gy), \tag{4.1}$$

are N-central, since they are respectively the graphs of $F_{x,y}, G_{x,y}$, and $v_{x,y}$; however the formally similar N-graph

$$(x,y) \longrightarrow (Gx,Fy), \tag{4.2}$$

is not N-central (There is no way to "get" from GA to FB, given only an arrow $A \longrightarrow B$). Similarly

$$I \longrightarrow (Fx,Fx),\quad I \longrightarrow (Gx,Gx),\quad I \longrightarrow (Fx,Gx),$$

are N-central, but $I \longrightarrow (Gx,Fx)$ is not.

Now define a <u>string</u> of length n = 1 to be an N-graph ,

$$\tau:I \longrightarrow (u_1,v_1), \tag{4.3}$$

where the variable u_1 has mate v_1 under τ. Here these "labels" v_1 and u_1 stand for any one of the symbols -, F-, G-, -', subject to the conditions that u_1 and v_1 have the same type and that if one is - so is the other. Similarly, define a string of length n > 1 to be an N-graph of the form

$$\sigma:(v_1,u_2)\otimes(v_2,u_3)\otimes\cdots\otimes(v_{n-1},u_n) \longrightarrow (u_1,v_n) \tag{4.4}$$

with each label v_i of the same type as u_i and mated to u_i by σ, $i = 1,\cdots,n$, and where $u_1 = -$ or $v_n = -$ implies that all variables are $-$. Finally, a block in a string (4.3) or (4.4) consists of a pair of indices $i \leq j$ with $u_i = G-$, $v_j = F-$, and all intermediate variables $v_i,\cdots,u_j$ (if any) equal to $-$. For example, blocks with $1 < i = j < n$ or $1 < i = j-1 < n-1$ are

$$(-',Gx)\otimes(Fx,Fy)\cdots, \qquad (-',Gx)\otimes(x,y)\otimes(Fy,Gz)\cdots .$$

The first case ($i=j$) has a covariant variable $G-$ with mate a contravariant $F-$, so might be called a "bad marriage". Again, the graphs (4.1) above are strings of length 2 with no blocks, while (4.2) is a string which has a block. We will wish strings without blocks. A string may be pictured as

It represents a sequence of morphisms which will be composable provided the mates (u_i & v_i) can be adjusted to be equal.

LEMMA 4.1. Every N-graph σ which is a string with no blocks is N-central, and is the graph of a unique N-central natural transformation s (in the category $\underline{V}_\#$).

PROOF. Suppose first that all variables are of type $-$; then each $v_i = u_i$. If $n = 1$, we may take s to be $j:I \longrightarrow (-,-)$, while if $n = 2$, we take s to be the identity $(x,y) \longrightarrow (x,y)$. For larger n, take s to be $n-2$ applications of composition Mc; by very familiar properties of the associative law for M, the same s results no matter the order in which the factors are composed. This is moreover the only such N-central s, since the only possible further variations would be inclusions of a factor I by b^{-1} followed later by $j:I \longrightarrow (A,A)$; since this must later be followed by an M, the rules (2.2) give the requisite uniqueness of s.

There remains the case when some variables are not of type -. Then by the definition of a string, the variables u_1 and v_n on the right cannot be of type -. In this case we obtain the desired natural s by iterated use of the composition M' in A', much as in the previous case, except that we first make preliminary adjustments, applying $k^\flat$ or k^* in order to make $u_i = v_i$ for every i. In case u_i and v_i are both -' no such adjustment is needed, and similarly if both are F- or both are G-. In case u_i is F- and v_i is G- the situation is in effect one in which we are given arrows $A' \longrightarrow FA$ and $GA \longrightarrow B'$; we can adjust to make these arrows composable by applying k, either as $k^\flat : (A',FA) \longrightarrow (A',GA)$ or as $k^*: (GA,B') \longrightarrow (FA,B')$. This choice between $k^\flat$ and k^* makes no difference, since a diagram chase from the definitions of k^* and $k^\flat$ via k shows that

$$\begin{array}{ccc}
(GA,B')\otimes(A',FA) & \xrightarrow{1\otimes k^\flat} & (GA,B')\otimes(A',GA) \\
\downarrow{\scriptstyle k^*\otimes 1} & & \downarrow{\scriptstyle M'} \\
(FA,B')\otimes(A',FA) & \xrightarrow[M']{} & (A',B')
\end{array}$$

always commutes. The case u_i = G- and v_i = F- cannot occur, as it would be a block with i = j (a "bad marriage"). There remain only cases when some but not all of the variables are -. Now if v_i is - in the shape (v_i, u_{i+1}), so is u_{i+1} (otherwise it is no shape), so consider the longest portion of the string containing v_i and with all variables equal to -. Writing the matching variables as A_i, this portion of the string has the form

$$(A_p,A_{p+1})\otimes(A_{p+1},A_{p+2})\otimes\cdots\otimes(A_{q-1},A_q) \tag{4.5}$$

with $p < q$; it is bordered on the left by (v_{p-1},u_p) with $u_p = FA_p$ or GA_p and on the right by (v_q,u_{q+1}), where $v_q=FA_q$ or GA_q. In case p = 1, the "bordering" term u_p is actually one which appears on the right side (u_1,v_n) of our string (4.4), and similarly for q = n. However, the discussion below will apply equally well to these extreme borders. Now the portion (4.5) of the string in case $\underline{V} = \underline{Set}$ amounts to giving a list of composable morphisms $A_p \longrightarrow A_{p+1} \longrightarrow A_{p+2} \longrightarrow \cdots \longrightarrow A_q$. Applying F, G, or $k:F \longrightarrow G$ we get the following commutative diagram

$$\begin{array}{ccccccc} FA_p & \longrightarrow & FA_{p+1} & \longrightarrow & \cdots & \longrightarrow & FA_q \\ \downarrow & & \downarrow & & & & \downarrow \\ GA_p & \longrightarrow & FA_{p+1} & \longrightarrow & \cdots & \longrightarrow & GA_q \end{array} ,$$

so that composition gives a unique morphism

$$FA_p \longrightarrow FA_q \quad , \quad FA_p \longrightarrow GA_q \quad , \quad GA_p \longrightarrow GA_q \quad ,$$

at our choice. We cannot get $GA_p \longrightarrow FA_q$, but the "no block" condition means that this will not be required. Translating this argument in the standard way to the $\underline{V}$-case, this means that application of F, G, k* and $k^\flat$, followed by the composition M', will yield a unique graph or natural transformation from the shape (4.5) to any one of the three shapes

$$(FA_p, FA_q), \quad (FA_p, GA_q), \quad (GA_p, GA_q) .$$

We choose that one which matches the given border $u_p = FA_p$ or GA_p and $v_q = FA_q$ or GA_q. As before (or use the argument of Lemma 4.2) the result is unique.

We next aim to show that every N-central morphism in $\underline{K}$ is a tensor product of strings. By definition, the N-central morphisms are formed from the central morphisms and the various morphisms

$$F,\ G,\ k^*,\ k^\flat;\ M,\ M';\ j,\ j',\ k, \tag{4.5}$$

by applying successively composition and tensor product. Since ⊗ is a bifunctor, we can rearrange the order, applying first ⊗ and then composition. Thus every N-central morphism is equal to a composite of central morphisms and "instances" of the morphisms (4.5); here by an <u>instance</u>, say, of F, we mean a morphism centrally conjugate to

$$1\otimes F_{AB}\otimes 1 : X\otimes(A,B)\otimes Y \longrightarrow X\otimes(FA,FB)\otimes Y$$

for some N-shapes X and Y. We say that this instance acts effectively only on the prime factor (A,B) of its domain.

LEMMA 4.2. Every N-central morphism in $\underline{K}$ is a composite of central morphisms and instances of the morphisms (4.5), taken in the order given in the list (4.5).

PROOF. First observe that to get this ordering we have listed k^* and $k^\flat$ among the "basic" morphisms (4.5), though they actually can be defined in terms of k and the others, as in (2.5) and (2.6).

Consider an arbitrary composite of instances of (4.5). Each such instance acts effectively on at most two of the prime factors of its domain, either on two factors, as for M and M', on just one, for F, G, k^*, or $k^\flat$, or on none, for j, j' and k. The order of application of two different instances in our composition can be interchanged ($\otimes$ is a functor) if the effective prime factors involved do not overlap; for example, any M may be interchanged with any M', any j with any j' or with any instance of k.

Now consider an instance of j in the composition; we assert that it can be moved (or dropped) so that it applies last in the composition. Indeed, j followed by F or G becomes j' by (2.3), j followed by an overlapping M will by (2.2) disappear (i.e., become central), and j cannot be followed by an overlapping k^*, $k^\flat$ or M'. On the other hand, j' followed by an overlapping M' disappears (again by the analog of (2.2)); j' followed by k^*,

$$I \xrightarrow{j'_{FA}} (FA,FA) \xrightarrow{k^*} (FA,GA)$$

is equal to k, by an easy application of the definition. These arguments move j, j' to the end of our composition.

Next, $k_A : I \longrightarrow (FA,GA)$ can't be followed by an overlapping F, G, k^*, $k^\flat$, or M. If it is followed by an overlapping M', this composite must be that used either in the definition (2.5) or the definition (2.6) of $k^\flat$ or k^*. Thus M' following k_A may be replaced by (a central morphism and) an instance of $k^\flat$ or k^*. This is the reason for listing $k^\flat$ and k^* in our basic list (4.5). With this argument, we have transported all (remaining) instances of k to the end of our composition.

Consider next instances of M or M'. If M is followed by an overlapping F or G, application of (2.3) replaces it by F (or G) followed by M'. If M' is followed by k^*, we may apply the commutative diagram

$$\begin{array}{ccc} (B',C')\otimes(GA,B') & \xrightarrow{M'} & (GA,C') \\ \Big\downarrow{\scriptstyle 1\otimes k^*_{A,B'}} & & \Big\downarrow{\scriptstyle k^*_{A,C'}} \\ (B',C')\otimes(FA,B') & \xrightarrow{M'} & (FA,C') \end{array} \quad ;$$

now M' is preceded by an instance of k^*. The same argument applies for M' followed by $k^\flat$. With these arguments we have achieved the order of composition indicated in (4.5), proving the lemma.

THEOREM 4.3. If T and S are N-shapes, then each N-central morphism $f:T \longrightarrow S$ of $\underset{\sim}{K}$ is centrally congruent to a morphism of the form

$$1\otimes h_1\otimes\cdots\otimes h_n : R\otimes T_1\otimes\cdots\otimes T_n \longrightarrow R\otimes S_1\otimes\cdots\otimes S_n \ ,$$

where the T_i, S_i and R are N-shapes, $1:R\longrightarrow R$ is the identity, and each N-graph Γh_i is a string $\sigma_i:T_i\longrightarrow S_i$ with no blocks, so that $\Gamma(1\otimes h_1\otimes\cdots\otimes h_n) = 1\otimes\sigma_1\otimes\cdots\otimes\sigma_n$.

PROOF. Consider an N-central morphism presented as a composite of instances (4.5) in the order now established. The instances of j, j', and k applied last in this composite each have a graph which is a string of length one and effective domain I; they yield some of the tensor factors σ_i as desired. Arrange the instances of M and M' (by central morphisms) into groups whose effective domains overlap. They clearly overlap in strings -- either strings of M, preceded by none of

F, G, k^* or $k^\flat$, or strings of M', possibly preceded by certain instances of

$$F_{AB}:(A,B)\longrightarrow(FA,FB), \qquad G_{AB}:(A,B)\longrightarrow(FA,GB),$$

$$k^*:(GA,B')\longrightarrow(FA,B'), \qquad k^\flat:(A',FB)\longrightarrow(A',GB).$$

Any easy induction shows that each such string σ_i has no blocks. Collecting these strings completes the proof of the theorem.

By combining the main results of this section we obtain a final theorem for N-central maps.

THEOREM 4.4. Let W be any closed naturality, and T, S any two N-shapes. Two N-central natural transformations f, $g:T\longrightarrow S$ in $\underline{V}_\#$ with the same N-graph are equal, and to any N-central N-graph $\xi:T\longrightarrow S$ in $\underline{V}_0$ there is an N-central morphism $f:T\longrightarrow S$ in $\underline{V}_\#$ with N-graph ξ.

PROOF. By Theorem 4.3 we may apply a central congruence to f (and also to g) so that the new f has the form $1 \otimes h_1 \otimes\cdots\otimes h_n$ with graph $1 \otimes \sigma_1 \otimes\cdots\otimes \sigma_n$ as described. Moreover, we can arrange that none of the prime factors of the identity $1:R\longrightarrow R$ are strings $(u,v)\longrightarrow(u,v)$ of length 1; then the list $\sigma_1,\cdots,\sigma_n$ includes all the strings in Γf. Now a further central congruence applied to g gives $g\longmapsto xgy = 1 \otimes k_1 \otimes\cdots\otimes k_m$ where each $\Gamma k_j = \tau_j$ is a string. But $\Gamma f = \Gamma g$, so $\Gamma(xgy)$ arises from Γf by a permutation of prime factors (Proposition 3.2) which is just a permutation ρ of the strings involved, so that $n = m$ and $\tau_j = \sigma_{\rho j}$. Therefore, k_j and $h_{\sigma j}$ have the same graph so must be equal by Lemma 4.1. This in turn implies $f = g$, as required.

The same theorem holds for W a monoidal naturality and T, S integral N-shapes. These theorems include a result analogous to Theorem 3.1 of [1], but they are stronger because they apply to N-central morphisms and <u>arbitrary</u> (not necessarily integral) N-shapes. A similar stronger theorem could have been stated (and proved) in the context of [1].

5. PROCESSES OF CONSTRUCTION

Again, $\underline{K}$ will be one of the closed categories $\underline{V}_0$ or $\underline{V}_\#$ for a given naturality W; the objects of $\underline{K}$ are N-shapes. We define the N-constructible morphisms of $\underline{K}$ to be the smallest class of morphisms containing all central morphisms, all central conjugates of morphisms in the class, all tensor products $f \otimes g$ of morphisms in the class, all adjuncts $\Pi f: A \longrightarrow [B,C]$ of morphisms $f: A \otimes B \longrightarrow C$ in the class, for $f: A \longrightarrow B$ and $g: C \otimes D \longrightarrow E$ in the class, all the composite morphisms

$$[B,C] \otimes A \otimes D \xrightarrow{<f> \otimes 1} C \otimes D \xrightarrow{g} S \quad , \tag{5.1}$$

where $<f>$ is defined to be the following composite with evaluation e:

$$[B,C] \otimes A \xrightarrow{1 \otimes f} [B,C] \otimes B \xrightarrow{e} C,$$

and finally (CM6) all unblocked strings, where we define an unblocked string to be an N-central morphism whose graph is a string with no blocks. The first five closure conditions are exactly the conditions CM1 - CM5 of [1]; we have added a sixth condition for strings which we call CM6.

Note especially that the central morphisms (CM1) and the N-central ones which are strings (CM6) play two quite different roles in this definition -- and in the subsequent proof. By Theorem 4.3, however, it does follow that every N-central morphism is N-constructible.

6. CONSTRUCTIBILITY OF ALLOWABLE MORPHISMS

The arguments of §6 in [1] now carry over, with "constructible" and "allowable" replaced by the new terms "N-constructible" and "N-allowable"; essentially by adding everywhere one new case "unblocked string". To start with, the proof of Proposition 6.2 of [1] extends trivially to give

PROPOSITION 6.2_N. Proposition 6.2 of [1] holds with one added case (v):

h is centrally conjugate to an unblocked string $s:X \longrightarrow Z$. Observe that, for such a string s, X and Z are N-integral and Z is prime.

To carry out inductions, we define for each N-shape a rank r, just as in [1] with the added proviso, R5, that $r(x,y) = 1$ for any term $(-,-)$ or (x,y); Lemma 6.3 still holds. In the crucial Proposition 6.4 (to show composites N-constructible) we are given h and k N-constructible and wish to show the composite

$$T\otimes U \xrightarrow{h\otimes 1} S\otimes U \xrightarrow{k} V \qquad (6.1)$$

N-constructible (here k is any morphism, not the k_A of the naturality). The new proof simply needs the following added cases, where one of h or k is of type "unblocked string".

Case 7. Both h and k are of type "unblocked string". Then $k(h \otimes 1)$ is obviously itself an unblocked string, and hence N-constructible. The final statement of our proposition asserts (for $\underline{K} = \underline{V}_0$) that the graphs of k and $h \otimes 1$ are compatible; this holds trivially in case 7: There are no "closed loops" because the string h mates no variables in its codomain S.

Case 8. h is of type unblocked string and k of type $\otimes$, so $k \cong f \otimes g$. Then $k(h \otimes 1)$ is centrally conjugate to a composite

$$X\otimes U \xrightarrow{s\otimes 1} Z\otimes U \cong A\otimes B \xrightarrow{f\otimes g} C\otimes D;$$

the central isomorphism in the middle associates the <u>prime</u> factor Z (we <u>use</u> this property of our strings) to a prime factor of A or of B, let us say of A. Then $A \cong Z \otimes A'$, and so $U \cong A' \otimes B$ and $k(h \otimes 1)$ is centrally conjugate to

$$X\otimes A'\otimes B \xrightarrow{s\otimes 1} Z\otimes A'\otimes B \cong A\otimes B \xrightarrow{f\otimes 1} C\otimes B \xrightarrow{1\otimes g} C\otimes D.$$

This has the form $(1 \otimes g)\circ(f' \otimes 1) = f' \otimes g$, where f' is

$$X\otimes A' \xrightarrow{s\otimes 1} Z\otimes A' \cong A \xrightarrow{f} C.$$

It is enough then to show f' N-constructible. Since g is given non-trivial, this follows from the induction assumption by an easy calculation of ranks.

Case 9. h is of type unblocked string and k is of type < >. Then our composite $k(h \otimes 1)$ is centrally conjugate to

$$X\otimes U \xrightarrow{s\otimes 1} Z\otimes U \cong [B,C]\otimes A\otimes D \xrightarrow{g(<f>\otimes 1)} V.$$

Now Z is both prime and N-integral, hence is associated by the middle congruence either to a factor of A or to a factor of D. There is a corresponding case subdivision, as follows.

Case 9a. Z associated to a factor of $A \cong Z \otimes A'$. Then also $U \cong [B,C] \otimes A' \otimes D$ and $h \otimes 1$ is conjugate to

$$[B,C]\otimes X\otimes A'\otimes D \xrightarrow{1\otimes s\otimes 1} [B,C]\otimes Z\otimes A'\otimes D,$$

so $k(h \otimes 1)$ is centrally conjugate to a composite $g\circ(<f'> \otimes 1)$ where f' is the morphism

$$X\otimes A' \xrightarrow{s\otimes 1} Z\otimes A' \cong A \xrightarrow{f} B.$$

Rank calculation and the induction assumption show this f' is N-constructible. The compatibility argument is similar.

Case 9b. Z associated to a factor of $D \cong Z \otimes D'$. Then $U \cong D' \otimes [B,C] \otimes A$ and $k(h \otimes 1)$ is centrally conjugate to

$$[B,C]\otimes A\otimes X\otimes D' \xrightarrow{<f>\otimes 1} C\otimes X\otimes D' \xrightarrow{g'} V$$

where g' is the morphism.

$$C\otimes X\otimes D' \xrightarrow{1\otimes s\otimes 1} C\otimes Z\otimes D' \cong C\otimes D \xrightarrow{g} V,$$

so that $g' \cong g(s \otimes 1)$; it is N-constructible by induction.

Case 10. h is of type π and k an unblocked string. In this case $h = \pi f: T \longrightarrow [B,C]$ for some $f: T \otimes B \longrightarrow C$ which is N-constructible. Therefore, $k(h \otimes 1)$ is

$$T \otimes U \xrightarrow{h \otimes 1} [B,C] \otimes U \cong X \xrightarrow{s} P \cong V.$$

Since X is an N-integral shape, which cannot have the prime factor [B,C], this case does not occur.

A table of all possible choices for h and k will show that we have now covered all cases, so that Proposition 6.2_N is proved.

The point of this proof is to show that the "cut elimination" process still applies in exactly the form used in our previous paper, essentially whenever there are additional types of constructible morphisms which have integral domain and prime and integral codomain--so that they do not "collide" with evaluation or adjunction.

THEOREM 6.5_N. The N-constructible morphisms of K are exactly the N-allowable ones.

The proof is like that of [1], using the observation of §5 that every N-central morphism is N-constructible and one added case AM6, which is easy.

Incidentally, in the previous proof [1], on line -9 of page 127, [g,f] is (three times) a misprint for [f,g].

THEOREM 2.2_N. If two N-graphs $\xi: T \longrightarrow S$ and $\eta: S \longrightarrow R$ are N-allowable, they are compatible.

This follows as before from the "compatible" clauses in Proposition 6.4_N. It shows that our "artificial" composition of non-compatibles is of no consequence.

7. THE COHERENCE THEOREM

In the present context, reduced N-shapes can be described as those shapes which are either I or else an iterated $\otimes$-product of prime factors alone (i.e., no factors I); as in Lemma 7.1 of [1], every N-shape is centrally conjugate to a reduced one. Lemma 7.2 carries over, because a morphism $s: X \longrightarrow Z$ of string type is conjugate to one with X and Z reduced.

THEOREM 2.1_N. Given N-shapes T and S, there is an algorithm for deciding when an N-graph $\xi:T \longrightarrow S$ is N-allowable.

PROOF. By the results of §6, we need only test when it is N-constructible. This we do by the algorithm of [1] plus the evident finite test that a graph is a string with no blocks.

LEMMA 7.3 of [1] applies to N-shapes without change. The notion of "proper" shapes is essential to

LEMMA 7.4_N. If $h:T \longrightarrow S$ is N-allowable in $\underline{K}$, with S constant and T proper, then T is constant.

This is proved inductively as before, with one new case, trivial because the codomain of a string is never constant.

LEMMA 7.5 carries over at once.

We now come to the essential sequence of Propositions of [1] which assert that the "construction" of N-allowable morphisms can always be done to follow a given construction (type $\otimes$, $\pitchfork$, $< >$, or "string") of their graphs -- provided certain shapes are proper.

Proposition 7.6 does this when $h:P \otimes Q \longrightarrow M \otimes N$ is given N-allowable with N-graph $\Gamma h = \xi \otimes \eta$ and P, Q, M, N are proper. The previous proof applies unchanged, since if $h \cong s$ of type string, Γh is not non-trivially of the form $\xi \otimes \eta$.

Proposition 7.7 holds as before.

Proposition 7.8_N considers a N-allowable morphism

$$h:[Q,M]\otimes P\otimes N \longrightarrow S \qquad (7.1)$$

between proper N-shapes, with [Q,M] not constant, and with Γh of the form $\eta(<\xi> \otimes 1)$; under supplementary hypotheses, one is to show that h itself has the form $q(<p> \otimes 1)$. The previous proof applies without change, since h cannot be of type string when its domain includes [Q,M].

We can now consider the main coherence theorem, already formulated in §2 as Theorem 2.4_N. We are to show that two N-allowable morphisms h, h':T $\longrightarrow$ S of the same N-graph between proper shapes are equal. But,by Theorem 6.5_N,h and h' are N-constructible, so we are reduced to the various five cases for h and for h'. To the previous proof we thus need add only one new case, when h is of type string, while h' is some other N-constructible with the same graph. But the visible form of a string graph clearly cannot be a graph for a morphism of type $\otimes$, or type π, or type < >, or a central morphism. Therefore h' is also of type string, and by Lemma 4.1 is the same string. The hard work was all done in [1] and §4 above.

REFERENCES

[1] G. M. Kelly and S. Mac Lane, Coherence in Closed Categories, J. Pure Appl. Alg. 1(1971) 97-140.

[2] G. M. Kelly, An abstract approach to Coherence, (in this volume).

[3] S. Eilenberg and G. M. Kelly, Closed Categories, in: Proc. Conf. on Categorical Algebra, La Jolla, 1965 (Springer-Verlag, 1966) pp. 421-562.

[4] G. Gentzen, Untersuchungen über das logische Schliessen I, II, Math. Z. 39(1934-1935) 176-210 and 405-431.

[5] J. Lambek, Deductive Systems and Categories I. Syntactic Calculus and Residuated Categories, Math. Systems Theory 2(1968) 287-318.

[6] S. Mac Lane, Natural Associativity and Commutativity, Rice University Studies 49(1963) 28-46.

[7] S. Mac Lane, Categories for the Working Mathematician, Springer-Verlag, Berlin, Heidelberg, New York 1972.

COHERENCE FOR DISTRIBUTIVITY

Miguel L. Laplaza
University of Chicago and
University of Puerto Rico at Mayaguez

Received November 5, 1971

INTRODUCTION

A familiar situation in category theory is given by a category, C, and two functors, $\otimes$, $\oplus$: $\underline{C} \times \underline{C} \longrightarrow \underline{C}$, that within natural isomorphisms are associative, commutative and such that $\otimes$ is distributive relative to $\oplus$. A coherence result for this situation is to characterize the diagrams whose commutativity is a consequence of the above structure and some suitable conditions on the natural isomorphisms. We are going to give a more precise description of this situation.

Let $\underline{C}$ be a category, $\otimes$, $\oplus$: $\underline{C} \times \underline{C} \longrightarrow \underline{C}$, two functors, U and N fixed objects of $\underline{C}$, called the unit and null objects. Suppose that we have natural isomorphisms,

$$\alpha_{A,B,C}: A\otimes(B\otimes C) \longrightarrow (A\otimes B)\otimes C, \quad \gamma_{A,B}: A\otimes B \longrightarrow B\otimes A,$$

$$\alpha'_{A,B,C}: A\oplus(B\oplus C) \longrightarrow (A\oplus B)\oplus C, \quad \gamma'_{A,B}: A\oplus B \longrightarrow B\oplus A,$$

$$\lambda_A: U\otimes A \longrightarrow A \quad , \quad \rho_A: A\otimes U \longrightarrow A, \tag{1}$$

$$\lambda'_A: N\oplus A \longrightarrow A \quad , \quad \rho'_A: A\oplus N \longrightarrow A,$$

$$\lambda^*_A: N\otimes A \longrightarrow N, \quad \rho^*_A: A\otimes N \longrightarrow N,$$

and natural monomorphisms,

$$\delta_{A,B,C}: A\otimes(B\oplus C) \longrightarrow (A\otimes B)\oplus(A\otimes C)$$
$$\delta^{\#}_{A,B,C}: (A\oplus B)\otimes C \longrightarrow (A\otimes C)\oplus(B\otimes C) \tag{2}$$

which are defined for any objects, A, B, C of $\underline{C}$.

A coherence result for the structure given to $\underline{C}$ by the family of isomorphisms, $\{\alpha_{A,B,C}: \lambda_A, \rho_A, \gamma_{A,B}\}$ was given by S. Mac Lane (see

[4] and [1]), and when we say that $\underline{C}$ is coherent for $\{\alpha_{A.B.C}, \lambda_A, \rho_A, \gamma_{A,B}\}$ or for $\{\alpha'_{A,B,C}, \lambda'_A, \rho'_A, \gamma'_{A,B}\}$ we want to refer to that result, although we are going to use the conditions in the form given by G. M. Kelly in [1].

We are going to give a coherence theorem for the above structure on $\underline{C}$, answering a question proposed in [5]. An announcement of this paper was given in [3]. Roughly speaking we intend to characterize the commutative diagrams that can be obtained by taking for vertices the combinations by $\otimes$ and $\oplus$ of objects of $\underline{C}$ and for arrows the combinations (also by $\otimes$ and $\oplus$) of the natural morphisms (1) and (2) with identities; to obtain a reasonable result we have to impose some conditions on $\underline{C}$ that are called the coherence conditions which hold in some usual situations.

The paper can be summarized in the following words: Let $X = \{x_1, x_2, \cdots, x_p, u, n\}$ be a set and construct the "free" category on the set X with functors $\otimes$ and $\oplus$ and with the natural morphisms (1) and (2); this is a category $\underline{C}(X)$ such that for any map, $m: X \longrightarrow \text{Ob}\, \underline{C}$, such that $m(u) = U$, $m(n) = N$, there exists one and only one functor, $\tilde{m}: \underline{C}(X) \longrightarrow \underline{C}$, extending the map m and preserving $\otimes$, $\oplus$ and the morphisms (1) and (2). The objects of $\underline{C}(X)$ will be the elements of the free algebra with two operations, $\{.,+\}$, over X and the arrows will be all the elements generated by . and + over formal symbols of type (1), (2) and identities. The coherence result states that if $\underline{C}$ satisfies the coherence conditions detailed in §1 and is regular (in the definition given later) then the image by $\tilde{m}$ of the set $\underline{C}(X)(a,b)$ has at most one element.

We have to remark that the construction of the category $\underline{C}(X)$ will be given almost completely, but we are not going to use the concept of "free" category given above.

From now on $\underline{C}$ will be a category with the structure given above, whose objects will be denoted by capital letters. We shall use the parenthesis with the usual conventions on sums and products and the symbols $\otimes$ will be omitted as often as possible.

The core of this work was done in the Department of Mathematics of the University of Chicago where the author spent one year as Post-doctoral Visitor and he wants to thank Professor S. Mac Lane for his illuminating direction and patient revision of the different versions of this paper.

§1. The Coherence Conditions

We will say that the category $\underline{C}$ is coherent when $\underline{C}$ is coherent in the sense of [1] separately for $\{\alpha, \gamma, \lambda, \rho\}$ and $\{\alpha', \gamma', \lambda', \rho'\}$ and the following types of diagrams are commutative for any vertices:

$$\begin{array}{ccc} A(B\oplus C) & \xrightarrow{\delta_{A,B,C}} & AB\oplus AC \\ \downarrow{\scriptstyle 1_A\cdot\gamma'_{B,C}} & & \downarrow{\scriptstyle \gamma'_{AB,AC}} \\ A(C\oplus B) & \xrightarrow{\delta_{A,C,B}} & AC\oplus AB \end{array} \quad , \qquad \text{(I)}$$

$$(\gamma_{A,C}\oplus\gamma_{B,C})\delta^{\#}_{A,B,C} = \delta_{C,A,B}\gamma_{A\oplus B,C} : (A\oplus B)C \longrightarrow CA\oplus CB \quad , \qquad \text{(II)}$$

$$\gamma'_{AC,BC}\delta^{\#}_{A,B,C} = \delta^{\#}_{B,A,C}(\gamma'_{A,B}\oplus 1_C) : (A\oplus B)C \longrightarrow BC\oplus AC \quad , \qquad \text{(III)}$$

$$\begin{array}{ccccc} [A\oplus(B\oplus C)]D & \xrightarrow{\delta^{\#}_{A,B\oplus C,D}} & AD\oplus(B\oplus C)D & \xrightarrow{1_{AD}\oplus\delta^{\#}_{B,C,D}} & AD\oplus(BD\oplus CD) \\ \downarrow{\scriptstyle \alpha'_{A,B,C}\cdot 1_D} & & & & \downarrow{\scriptstyle \alpha'_{AD,BD,CD}} \\ [(A\oplus B)\oplus C]D & \xrightarrow{\delta^{\#}_{A\oplus B,C,D}} & (A\oplus B)D\oplus CD & \xrightarrow{\delta^{\#}_{A,B,D}\oplus 1_{CD}} & (AD\oplus BD)\oplus CD \end{array} \quad , \qquad \text{(IV)}$$

$$\begin{array}{ccccc}
A[B\oplus(C\oplus D)] & \xrightarrow{\delta_{A,B,C\oplus D}} & AB\oplus A(C\oplus D) & \xrightarrow{1_{AB}\oplus\delta_{A,C,D}} & AB\oplus(AC\oplus AD) \\
\downarrow{\scriptstyle 1_A\cdot\alpha'_{B,C,D}} & & & & \downarrow{\scriptstyle \alpha'_{AB,AC,AD}} \\
A[(B\oplus C)\oplus D] & \xrightarrow{\delta_{A,B\oplus C,D}} & A(B\oplus C)\oplus AD & \xrightarrow{\delta_{A,B,C}\oplus 1_{AD}} & (AB\oplus AC)\oplus AD
\end{array} \quad , \qquad \text{(V)}$$

$$\begin{array}{ccccc}
A[B(C\oplus D)] & \xrightarrow{1_A\cdot\delta_{B,C,D}} & A(BC\oplus BD) & \xrightarrow{\delta_{A,BC,BD}} & A(BC)\oplus A(BD) \\
\downarrow{\scriptstyle \alpha_{A,B,C\oplus D}} & & & & \downarrow{\scriptstyle \alpha_{A,B,C}\oplus\alpha_{A,B,D}} \\
(AB)(C\oplus D) & & \xrightarrow{\qquad\delta_{AB,C,D}\qquad} & & (AB)C\oplus(AB)D
\end{array} \qquad \text{(VI)}$$

$$\begin{array}{ccccc}
(A\oplus B)(CD) & & \xrightarrow{\qquad\delta^{\#}_{A,B,CD}\qquad} & & A(CD)\oplus B(CD) \\
\downarrow{\scriptstyle \alpha_{A\oplus B,C,D}} & & & & \downarrow{\scriptstyle \alpha_{A,C,D}\oplus\alpha_{B,C,D}} \\
[(A\oplus B)C]D & \xrightarrow{\delta^{\#}_{A,B,C}\cdot 1_D} & (AC\oplus BC)D & \xrightarrow{\delta^{\#}_{AC,BC,D}} & (AC)D\oplus(BC)D
\end{array} \quad , \qquad \text{(VII)}$$

$$\begin{array}{ccccc}
A[(B\oplus C)D] & \xrightarrow{1_A\cdot\delta^{\#}_{B,C,D}} & A(BD\oplus CD) & \xrightarrow{\delta_{A,BD,CD}} & A(BD)\oplus A(CD) \\
\downarrow{\scriptstyle \alpha_{A,B\oplus C,D}} & & & & \downarrow{\scriptstyle \alpha_{A,B,D}\oplus\alpha_{A,C,D}} \\
[A(B\oplus C)]D & \xrightarrow{\delta_{A,B,C}\cdot D} & (AB\oplus AC)D & \xrightarrow{\delta^{\#}_{AB,AC,D}} & (AB)D\oplus(AC)D
\end{array} \quad . \qquad \text{(VIII)}$$

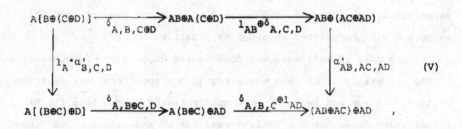

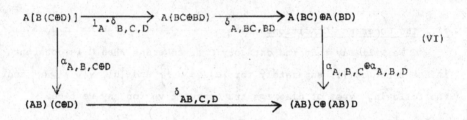

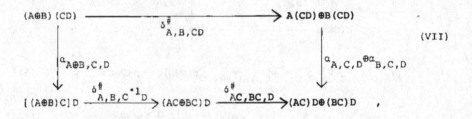

$$
\begin{array}{ccc}
 & A(C\oplus D)\oplus B(C\oplus D) \xrightarrow{\delta_{A,C,D}\oplus\delta_{B,C,D}} & (AC\oplus AD)\oplus(BC\oplus BD) \\
 & & \downarrow \alpha'_{AC\oplus AD,BC,BD} \\
 & & [(AC\oplus AD)\oplus BC]\oplus BD \\
 \nearrow \delta^{\#}_{A,B,C\oplus D} & & \downarrow \alpha'^{-1}_{AC,AD,BC}\oplus 1_{BD} \\
 & & [AC\oplus(AD\oplus BC)]\oplus BD \\
(A\oplus B)(C\oplus D) & & \downarrow (1_{AC}\oplus\gamma'_{AD,BC})\oplus 1_{BD} \\
 & & [AC\oplus(BC\oplus AD)]\oplus BD \\
 \searrow \delta_{A\oplus B,C,D} & & \downarrow \alpha'_{AC,BC,AD}\oplus 1_{BD} \\
 & & [(AC\oplus BC)\oplus AD]\oplus BD \\
 & & \uparrow \alpha'_{AC\oplus BC,AD,BD} \\
 & (A\oplus B)C\oplus(A\oplus B)D \xrightarrow{\delta^{\#}_{A,B,C}\oplus\delta^{\#}_{A,B,D}} & (AC\oplus BC)\oplus(AD\oplus BD)
\end{array} \quad , \qquad \text{(IX)}
$$

$$
\lambda^*_N = \rho^*_N : N\,N \longrightarrow N \quad , \qquad \text{(X)}
$$

$$
\begin{array}{ccc}
N(A\oplus B) & \xrightarrow{\delta_{N,A,B}} & NA\oplus NB \\
\downarrow \lambda^*_{A\oplus B} & & \downarrow \lambda^*_A\oplus\lambda^*_B \\
N & \xleftarrow{\lambda'_N} & N\oplus N
\end{array} \qquad \text{(XI)}
$$

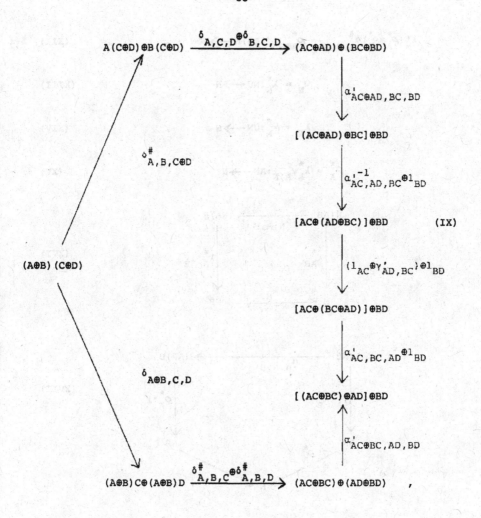

$$\lambda'_N(\rho^*_A\oplus\rho^*_B)\delta^{\#}_{A,B,N} = \rho^*_{A\oplus B}:(A\oplus B)N \longrightarrow N\oplus N \quad , \tag{XII}$$

$$\rho_N = \lambda^*_N : NU \longrightarrow N \quad , \tag{XIII}$$

$$\lambda_N = \rho^*_N : UN \longrightarrow N \quad , \tag{XIV}$$

$$\rho^*_A = \lambda^*_A \gamma'_{A,N} : AN \longrightarrow N \quad , \tag{XV}$$

$$\begin{array}{ccc}
N(AB) & \xrightarrow{\alpha_{N,A,B}} & (NA)B \\
\big\downarrow{\scriptstyle \lambda^*_{AB}} & & \big\downarrow{\scriptstyle \lambda^*_A\cdot 1_B} \\
N & \xleftarrow{\lambda^*_B} & NB
\end{array} \quad , \tag{XVI}$$

$$\begin{array}{ccc}
A(NB) & \xrightarrow{\alpha_{A,N,B}} & (AN)B \\
\big\downarrow{\scriptstyle 1_A\cdot\lambda^*_B} & & \big\downarrow{\scriptstyle \rho^*_A\cdot 1_B} \\
AN & & NB \\
& \searrow{\scriptstyle \rho^*_A} \quad {\scriptstyle \lambda^*_B}\swarrow & \\
& N &
\end{array} \quad , \tag{XVII}$$

$$\rho^*_{AB}\alpha_{A,B,N} = \rho^*_A(1_A\cdot\rho^*_B) : A(BN) \longrightarrow N \quad , \tag{XVIII}$$

$$\begin{array}{ccc}
A(N\oplus B) & \xrightarrow{\delta_{A,N,B}} & AN\oplus AB \\
\big\downarrow{\scriptstyle 1_A\cdot\lambda'_B} & & \big\downarrow{\scriptstyle \rho^*_A\oplus 1_{AB}} \\
AB & \xleftarrow{\lambda'_{AB}} & N\oplus AB
\end{array} \quad , \tag{XIX}$$

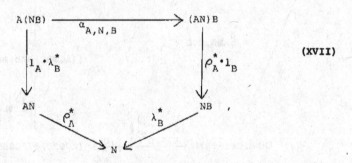

$$\lambda'_{BA}(\lambda^*_A \oplus 1_{BA})\delta^{\#}_{N,B,A} = \lambda'_B \otimes 1_A : (N \oplus B)A \longrightarrow BA \quad , \qquad \text{(XX)}$$

$$\rho'_{AB}(1_{AB} \oplus \rho^*_A)\delta_{A,B,N} = 1_A \oplus \rho'_B : A(B \oplus N) \longrightarrow AB \quad , \qquad \text{(XXI)}$$

$$\rho'_{AB}(1_{AB} \oplus \lambda^*_B)\delta^{\#}_{A,N,B} = \rho'_A \otimes 1_B : (A \oplus N)B \longrightarrow AB \quad , \qquad \text{(XXII)}$$

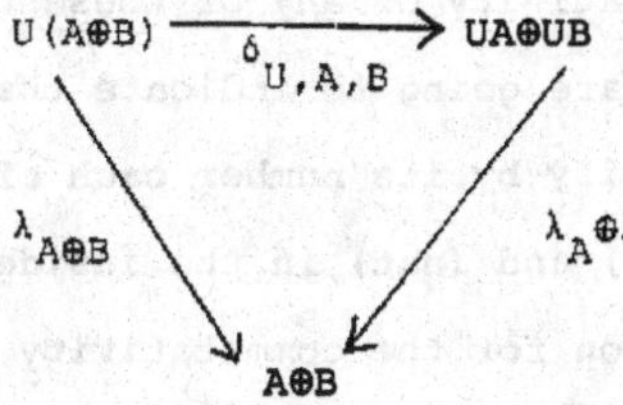

(XXIII)

$$(\rho_A \oplus \rho_B)\delta_{A,B,U} = \rho_{A\oplus B} : (A \oplus B)U \longrightarrow A \oplus B \quad , \qquad \text{(XXIV)}$$

The commutativity of some types of diagrams imply the commutativity of others, and we are going to indicate some of those relations. A detailed study of the minimal conditions assuring the coherence of C for $\{\alpha, \gamma, \lambda, \rho\}$ or $\{\alpha', \gamma', \lambda', \rho'\}$ is contained in [1].

We will prove the following set of relations, in which the number of the diagram denotes the condition of commutativity of all the diagrams of that type:

1) (II) $\Longrightarrow$ ((I) $\Longleftrightarrow$ (III)),
2) (II) $\Longrightarrow$ ((IV) $\Longleftrightarrow$ (V)),
3) Coherence of C for $\{\alpha, \gamma, \lambda, \rho\} \wedge$ (II) $\Longrightarrow$ ((VI) $\Longleftrightarrow$ (VII)),
4) Coherence of C for $\{\alpha, \gamma, \lambda, \rho\} \wedge$ (II) $\Longrightarrow$ ((VI) $\Longrightarrow$ (VIII)),
5) Coherence of C for $\{\alpha, \gamma, \lambda, \rho\} \wedge$ (XV) $\Longrightarrow$ ((XIII) $\Longleftrightarrow$ (XIV)),
6) (II) $\wedge$ (XV) $\Longrightarrow$ ((XI) $\Longleftrightarrow$ (XII)) ,
7) Coherence of C for $\{\alpha', \gamma', \lambda', \rho'\} \wedge$ (XV) $\Longrightarrow$
$\Longrightarrow$ Any two of $\{$(XVI), (XVII), (XVIII)$\}$ imply the other ,

8) (XV) $\wedge$ (I) $\wedge$ (II) $\Longrightarrow$ Each one of $\{$(XIX), (XX), (XXI), (XXII)$\}$ implies the others,

9) Coherence of $\underline{C}$ for $\{\alpha, \gamma, \lambda, \rho\}$ $\wedge$ (II) $\Longrightarrow$ ((XXIII) $\Longleftrightarrow$ (XXIV)).

The proof of all the above relations uses the same method: the construction of a diagram in which the commutativity of all the subdiagrams with the exception of two follows from the hypothesis of the relation so that the commutativity of any of these two diagrams are equivalent conditions. We are going to indicate the construction of these diagrams and to identify by its number each of the subdiagrams involved. The symbols (coh) and (nat) in the inside of a subdiagram will indicate that the reason for the commutativity is the coherence of $\underline{C}$ for $\{\alpha, \gamma, \lambda, \rho\}$ or the naturality of the elements involved in the construction of the subdiagram.

Proof of 1): It is given by the following diagram in which the outside is of type (I)

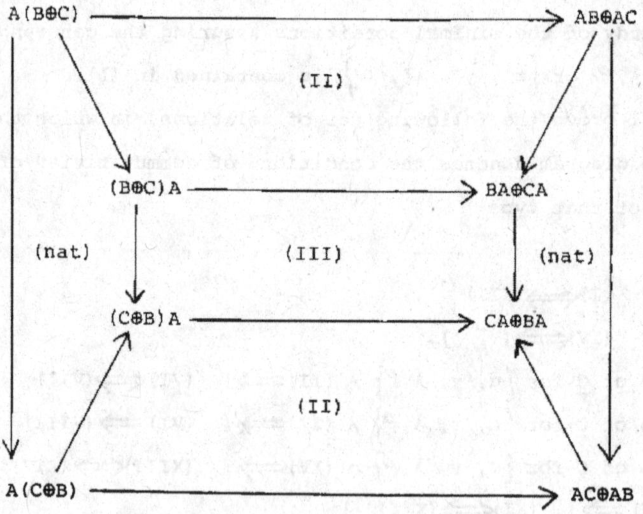

.

Proof of 2): It is given by the following diagram in which the outside is (V)

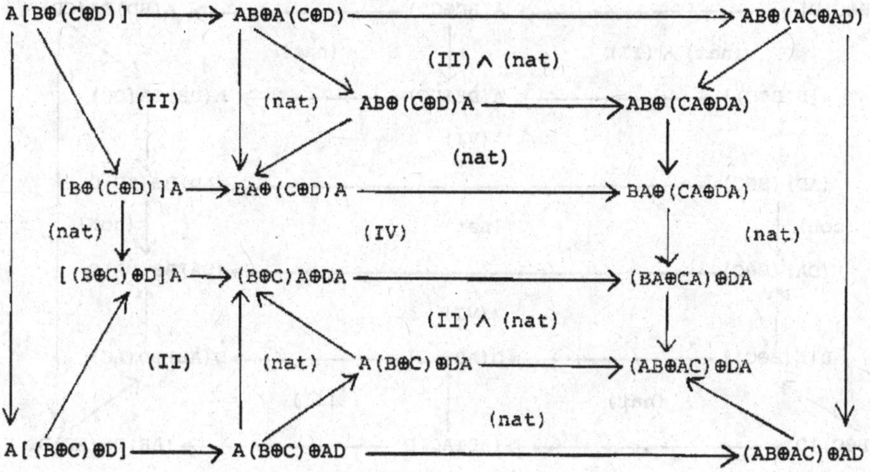

Proof of 3): It is given by the following diagram in which the outside is (VI)

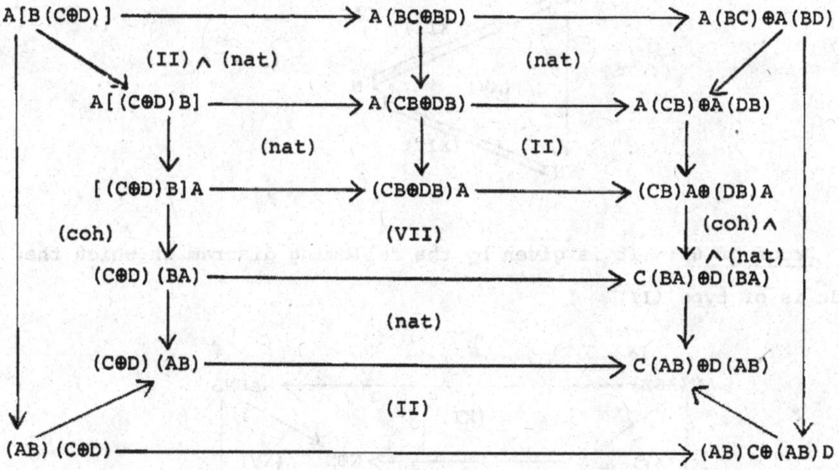

.

Proof of 4): It is given by the following diagram in which the outside is (VIII)

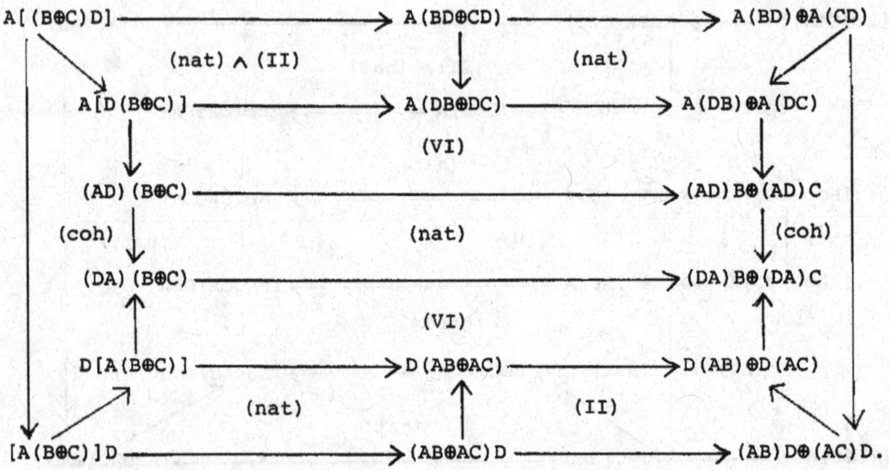

Proof of 5): It is given by the following diagram in which the outside diagram commutes by the coherence of C for $\{\alpha, \gamma, \lambda, \rho\}$

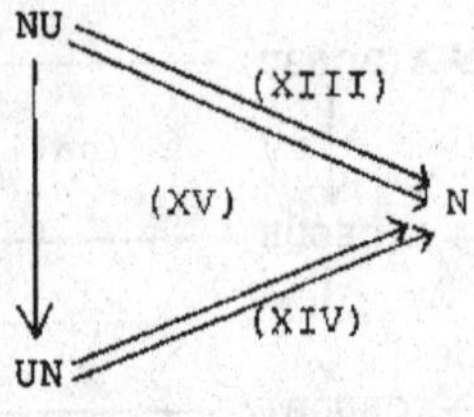

Proof of 6): It is given by the following diagram in which the outside is of type (II)

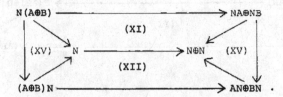

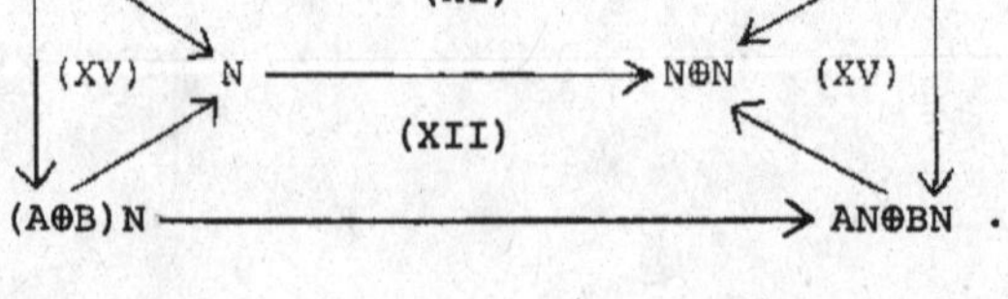

<u>Proof of 7)</u>: **It is given by the following diagram in which the outside is commutative by the coherence of <u>C</u> for** $\{\alpha, \lambda, \rho, \gamma\}$

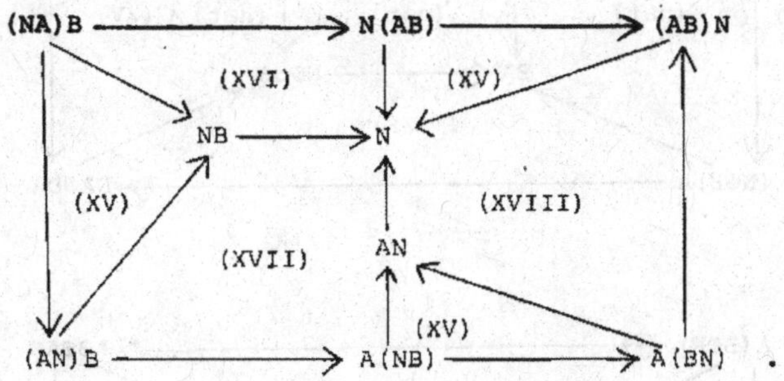

<u>Proof of 8)</u>: **It is given by the following diagrams in which the outside are of type (I), (II), and (II) respectively**

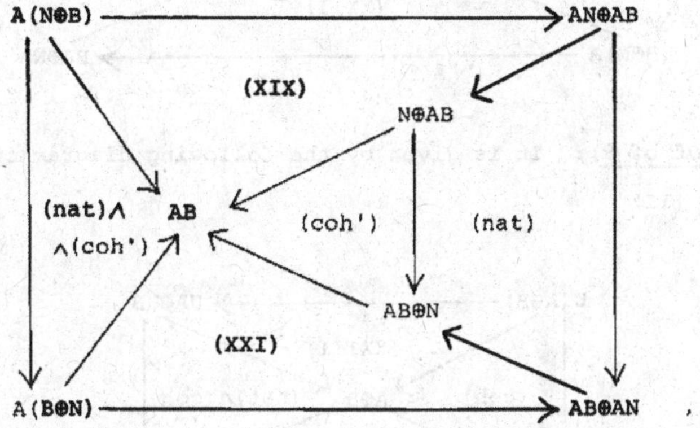

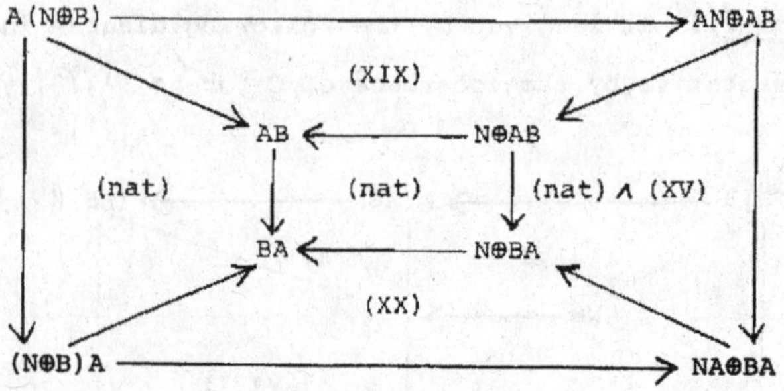

,

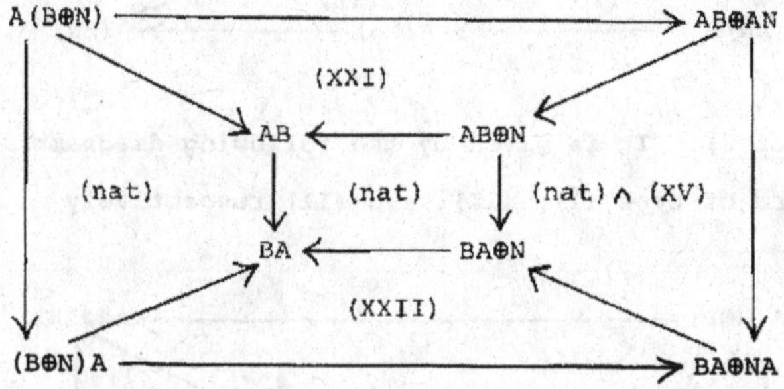

.

<u>Proof of 9)</u>: It is given by the following diagram in which the outside is (II)

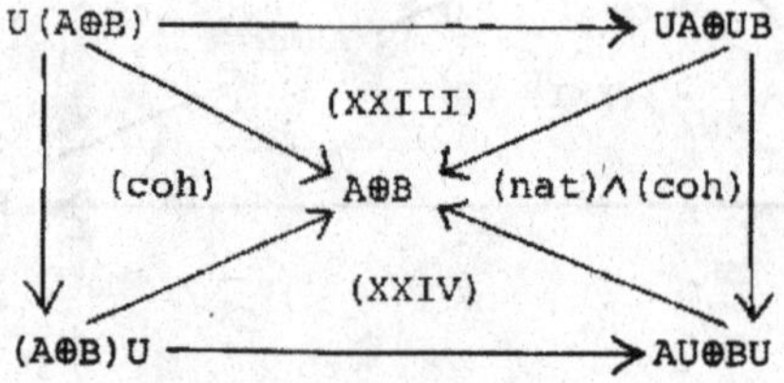

.

An immediate consequence of the above relations is that for $\underline{C}$ to be coherent it is sufficient to check that $\underline{C}$ satisfies the following conditions:

1) $\underline{C}$ is coherent for $\{\alpha, \gamma, \lambda, \rho\}$ and for $\{\alpha', \gamma', \lambda', \rho'\}$.

2) All the diagrams of type (II), (IX), (X) and (XV) are commutative.

3) For one type contained in each one of the sets, $\{(I), (III)\}$, $\{(IV), (V)\}$, $\{(VI), (VII)\}$, $\{(XI), (XII)\}$, $\{(XIII), (XIV)\}$, $\{(XIX), (XX), (XXI), (XXII)\}$, $\{(XXIII), (XXIV)\}$, all the diagrams are commutative.

3) For two of the types contained in $\{(XVI), (XVII), (XVIII)\}$ all the diagrams are commutative.

§2. Definition and evaluation of the paths: Formulation of the coherence problem

Let X be the set $\{x_1, x_2, \cdots, x_p, n, u\}$, $\underline{A}$ the free $\{+,.\}$-algebra over X and $\underline{G}$ the graph consisting of all the following formal symbols, for x, y, z $\in \underline{A}$,

$$\alpha_{x,y,z}: x(yz) \longrightarrow (xy)z \quad , \quad \alpha'_{x,y,z}: x + (y + z) \longrightarrow (x + y) + z,$$

$$\lambda_x: ux \longrightarrow x \quad , \quad \lambda'_x: n + x \longrightarrow x$$

$$\rho_x: xu \longrightarrow x \quad , \quad \rho'_x: x + n \longrightarrow x \ ,$$

$$\gamma_{x,y}: xy \longrightarrow yx \quad , \quad \gamma'_{x,y}: x + y \longrightarrow y + x \ ,$$

$$\lambda^*_x: nx \longrightarrow n \ ,$$

$$\rho^*_x: xn \longrightarrow n \ ,$$

their formal inverses, indicated by the upper index -1, and

$$\delta_{x,y,z}: x(y + z) \longrightarrow xy + xz \ ,$$

$$\delta^{\#}_{x,y,z}: (x + y)z \longrightarrow xz + yz \ ,$$

$$1_x: x \longrightarrow x .$$

Note that we use the symbol $\longrightarrow$ to indicate the edges of the graph to distinguish them from the arrows of the category denoted by $\rightarrow$.

Let $\underline{H}$ be the free $\{+,1\}$-algebra over the edges of $\underline{G}$ and take on $\underline{H}$ the unique extension of the graph structure of $\underline{G}$ in which the projections are $\{+,.\}$-morphisms. An element of $\underline{H}$ is an <u>instantiation</u> if, with at most one exception, only elements of $\underline{G}$ of type 1_x are involved in its expression: the elements involving only elements of $\underline{G}$ of type 1_x are called instantiations of identities or simply identities. We will denote by $\underline{T}$ the graph consisting of all the instantiations of $\underline{H}$. We can define now the <u>paths</u> as the sequences,

$$y_1 \xrightarrow{\Pi_1} y_2 \xrightarrow{\Pi_2} \cdots \xrightarrow{\Pi_{m-1}} y_m \ ,$$

where $\Pi_i \in \underline{T}$. We can speak of the existence of diagrams involving elements of $\underline{T}$, but not yet of the commutativity of such diagrams because the product of edges of $\underline{T}$ is not (and will not be) defined.

Fix now p objects, $0_1, 0_2, \cdots, 0_p$ in $\underline{C}$ and let $g:\underline{T} \longrightarrow \underline{C}$ be the morphism of graphs defined on the vertices by the conditions, i) $gu = U$, $gn = N$, $gx_i = 0_i$, for $i = 1, 2, \cdots, p$, ii) $g(x + y) = gx \oplus gy$, $g(xy) = gx \otimes gy$, for $x,y \in \underline{A}$; on $\underline{G}$ by taking each formal symbol onto the arrow of $\underline{C}$ determined replacing each subscript by its image by g and such that for $x, y \in \underline{T}$, $g(x + y) = gx \oplus gy$, $g(xy) = gx \otimes gy$. This definition depends upon the 0_i and allows us to define the <u>value</u> of a path as the product of the value of the steps and to define that a diagram with elements in $\underline{T}$ is commutative if any two paths contained in the diagram and with the same origin and end have the same value.

An ideal coherence result would state that if $\underline{C}$ is coherent in the sense of §1, then for any choice of the 0_i any diagram of elements of $\underline{T}$ is commutative, that is, for any choice of the 0_i the value of any path only depends upon the origin and the end of the path. But this is not true in some simple cases; for instance if $\underline{C}$ is the category of unitary modules over a commutative ring, $\otimes$ the tensor product, $\oplus$ the direct sum and if 0_1 is not the null module, then the value of $1_{x_1+x_1}:x_1 + x_1 \longrightarrow x_1 + x_1$ is the identity map of $0_1 \oplus 0_1$ and the value of $\gamma'_{x_1,x_1}:x_1 + x_1 \longrightarrow x_1 + x_1$ is the map defined by $\langle a,b\rangle \longrightarrow \langle b,a\rangle$ that is not the identiry. In this sense the coherence problem has a negative answer but we are going to prove that it is sufficient to impose a reasonable restriction on the vertices of the diagrams to get a coherence result that holds for any choice of the 0_i.

Note that the free category generated by the graph G would be the free category C(X) referred to in the Introduction.

We shall use the symbol $\xrightarrow{\theta}$ to indicate paths with steps in T. The expression a $\xrightarrow{\theta}$ b will denote the existence of a path from a to b.

§3. Regularity and some preliminary concepts

We shall indicate by N the set of natural numbers and by $S^{[N]}$ the set of all finite sequences of elements of S. In general, we shall represent the elements of $S^{[N]}$ by putting into parenthesis the sequence of the elements, identifying the elements of S with the sequences of $S^{[N]}$ with only one element.

All the definitions included in this part, with the exception of the concept of regularity, are auxiliary tools to be used in the proof of the propositions.

The rank of the elements of A is defined by means of the map, rank:A $\longrightarrow$ N, uniquely determined by the following conditions,

i) For $x \in X$, rank $x = 2$,

ii) For a, b $\in$ A, rank $(a + b)$ = rank(ab) = rank a + rank b.

The size, siz:A $\longrightarrow$ N, is defined by the conditions,

i) For $x \in X$, siz $x = 2$,

ii) For a,b $\in$ A, siz(a + b) = siz a + siz b, siz(ab) = (siz a)(siz b).

It is very easy to prove that for any element, y, of A,

$$\text{rank } y \leq \text{siz } y,$$

and that rank y = siz y, if and only if y is the sum of elements of that are products of elements of S.

The norm, $|\,|$:A $\longrightarrow$ N, is uniquely defined by the conditions,

i) For $x \in X$, $|x| = 1$,

ii) For a, b $\in \underline{A}$, $|a + b| = |ab| = |a| + |b|$.

The additive decomposition, Adec: $\underline{A} \longrightarrow \underline{A}^{[N]}$, is defined by the conditions,

i) For $x \in X$, Adec x = x,

ii) For y, z $\in \underline{A}$, Adec(yz) = yz,

iii) If Adec a = $(a_1,\cdots,a_r)$, Adec b = $(b_1,\cdots,b_s)$, then

$$\text{Adec}(a + b) = (a_1,\cdots,a_r,b_1,\cdots,b_s).$$

In a similar way, the multiplicative decomposition, Mdec: $\underline{A} \longrightarrow \underline{A}^{[N]}$, is defined by the conditions,

i) For $x \in X$, Mdec x = x,

ii) For y, z $\in \underline{A}$, Mdec(y + z) = y + z,

iii) If Mdec a = $(a_1,\cdots,a_r)$, Mdec b = $(b_1,\cdots,b_s)$, then

Mdec(ab) = $(a_1,\cdots,a_r,b_1,\cdots,b_s)$.

The additive pattern of the top, Apt: $\underline{A} \longrightarrow \underline{A}$, is defined by the conditions,

i) For x, y $\in \underline{A}$, Apt(x + y) = Apt x + Apt y,

ii) For x $\in \underline{A}$, if Adec x = x, then, Apt x = x_1.

In a similar way, the multiplicative pattern of the top, Mpt: $\underline{A} \longrightarrow \underline{A}$, is defined by the conditions,

i) For x,y $\in \underline{A}$, Mpt(xy) = (Mpt x)(Mpt y),

ii) For x $\in \underline{A}$, if Mdec x = x, then, Mpt x = x_1.

Proposition 1

For any elements a and b of $\underline{A}$ we have the following relations:

i) Apt a = Apt b $\wedge$ Adec a = Adec b $\Longrightarrow$ a = b.

ii) Mpt a = Mpt b $\wedge$ Mdec a = Mdec b $\Longrightarrow$ a = b.

Proof:

It will be sufficient to prove one of the relations, say i). If Apt a = Apt b = x_1, then, a = Adec a = Adec b = b, and the relation is proved. Suppose now that,

$$\text{Apt } a = \text{Apt } b = x + y ,$$

$$\text{Adec } a = \text{Adec } b = (c_1, \cdots, c_t).$$

Then it is immediate that if $|x| = r$, then, $a = a' + a''$, $b = b' + b''$ with

$$\text{Apt } a' = \text{Apt } b' = x, \quad \text{Apt } a'' = \text{Apt } b'' = y,$$

$$\text{Adec } a' = \text{Adec } b' = (c_1, \cdots, c_r)$$

$$\text{Adec } a'' = \text{Adec } b'' = (c_{r+1}, \cdots, c_t).$$

From these facts, the proof of the proposition by induction on $|\text{Apt } a|$ is immediate.

Let $\underline{A}^*$ be the free $\{+,.\}$-algebra over X, with associativity and commutativity for . and +, distributivity of . relatively to +, null element n, identity element u, and the additional condition, $na = an = n$ for $a \in \underline{A}^*$. $\underline{A}^*$ is a strict algebra and the identity map of X defines a $\{+,.\}$-morphism, called the support, $\text{Supp}: \underline{A} \longrightarrow \underline{A}^*$. That means that the support is defined by the following conditions:

i) If $x \in X$, $\text{Supp } x = x \in \underline{A}^*$,

ii) If $x,y \in \underline{A}$, $\text{Supp}(x + y) = \text{Supp } x + \text{Supp } y$,

iii) If $x,y \in \underline{A}$, $\text{Supp}(xy) = (\text{Supp } x)(\text{Supp } y)$.

An element a of $\underline{A}$ is defined to be <u>regular</u> if Supp a can be expressed as a sum of different elements of $\underline{A}^*$ each of which is a product of different elements of X. In any concrete case this definition can be easily checked, but we shall present later (Proposition 3) another simple case in which the regularity of an element can immediately be asserted.

<u>Proposition 2</u>.

Suppose $a \xrightarrow{\theta} b$, that is, assume the existence of a path from a to b. Then, a is regular if and only if b is regular.

<u>Proof</u>:

It is easy to prove that, $a \longrightarrow b \Longrightarrow \text{Supp } a = \text{Supp } b$, and hence,

$a \xrightarrow{\theta} b \Longrightarrow$ Supp a = Supp b, and this relation immediately proves the proposition.

Define the <u>elemental components</u>, Ecomp: $\underline{A} \longrightarrow \mathcal{P}(X)$, the power set of X, by the conditions:

i) If $x \in X$, Ecomp $x = \{x\}$,

ii) For a, b $\in \underline{A}$, Ecomp(x + y) = Ecomp(xy) = Ecomp x $\cup$ Ecomp y.

<u>PROPOSITION 3</u>

Suppose that a is an element of $\underline{A}$ such that any element of X appears at most once in the expression of a. Then, a is regular.

<u>Proof</u>.

The first thing to prove is that if x and y are regular elements of $\underline{A}$ such that, Ecomp x $\cap$ Ecomp y = $\emptyset$, then xy and x + y are also regular elements and this is routine.This fact allows us to prove immediately the proposition by induction on $|a|$, because if a = xy or a = x + y, then, the proposition hypothesis implies, Ecomp x $\cap$ Ecomp y = $\emptyset$.

Observe that if a is not a regular element, it is possible to find a path, $a \xrightarrow{\theta} b$, where b involves a situation of type x + x or xx: as it has been noted in the counterexample included in §2, this type of element originates an "incoherent" diagram in some usual cases.

<u>§4. The concept of reduction</u>

Let a be an element of $\underline{A}$. A <u>reduction</u> of a is a path $a \xrightarrow{\theta} a'$ such that,

i) Every step in the path is an instantiation of λ^*, ρ^*, λ', ρ' or an identity.

ii) a' = n or there is no occurrence of n in the expression of a'.

Note that the condition ii) is equivalent to say that a' is not the origin of an instantiation of λ^*, ρ^*, λ', or ρ'. Intuitively speaking a reduction of a is any path obtained by elimination of n in a by means of λ^*, ρ^*, λ' and ρ'.

PROPOSITION 4

Let a be an element of $\underline{A}$. Then, there exists a reduction $a \xrightarrow{\theta} a'$ of a, a' is uniquely determined by a and if $\underline{C}$ is coherent the value of the reduction is unique.

Proof:

The proof of the existence of a reduction of a can be done immediately by induction on rank a.

For the proof of the uniqueness of a' we have to state some preliminary relations:

1) Supp a = n $\Longleftrightarrow$ a' = n.
   It is clear that Supp a = Supp a'; hence if Supp a = n, then, a' = n because otherwise the expression of a' and also of Supp a' would involve no occurrence of n.

2) If $a = a_1 + a_2$ and $a_1 \xrightarrow{\theta} a_1'$, $a_2 \xrightarrow{\theta} a_2'$ are reductions we have,

$$a_1' \neq n \wedge a_2' \neq n \Longrightarrow a' = a_1' + a_2' \quad ,$$

$$a_1' = n \wedge a_2' \neq n \Longrightarrow a' = a_2' \quad ,$$

$$a_1' \neq n \wedge a_s' = n \Longrightarrow a' = a_1' \quad .$$

The proof of the above assertion can be done very easily by induction on rank a.

3) If $a = a_1 a_2$ and $a_1 \xrightarrow{\theta} a_2'$, $a_2 \xrightarrow{\theta} a_2'$ are reductions we have,

$$a_1' \neq n \wedge a_2' \neq n \Longrightarrow a' = a_1' a_2' \quad .$$

The proof is similar to the proof of 2).

The above three assertions allow us to prove immediately the uniqueness of a' by induction on $|a|$.

Suppose that $a \longrightarrow b$ and $a \longrightarrow c$ are instantiations of λ^*, ρ^*, λ' or ρ' and that $\underline{C}$ is coherent; as a preliminary step to end the proof of the proposition we need to prove the existence of a commutative diagram of type

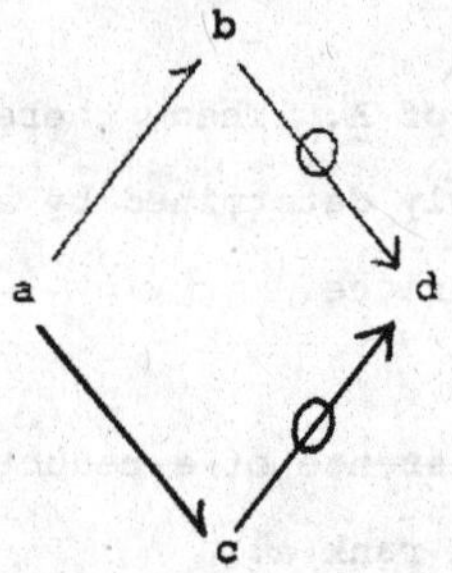

such that any step in $b \xrightarrow{\theta} d$ and $c \xrightarrow{\theta} d$ is an identity or an instantiation of λ^*, ρ^*, λ', or ρ'. The proof is a routine induction on $|a|$ outlined by the following diagrams (and other analogous diagrams).

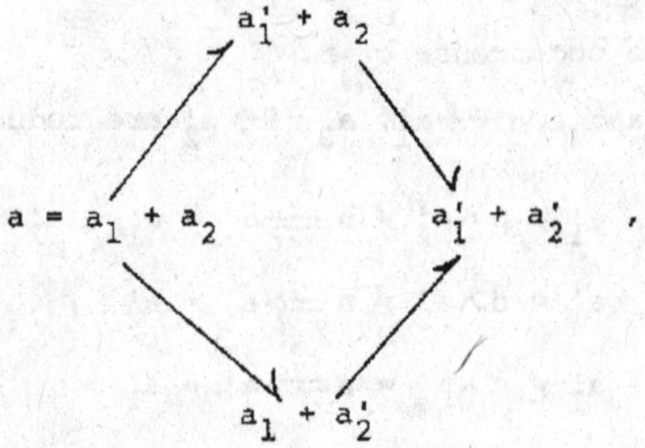

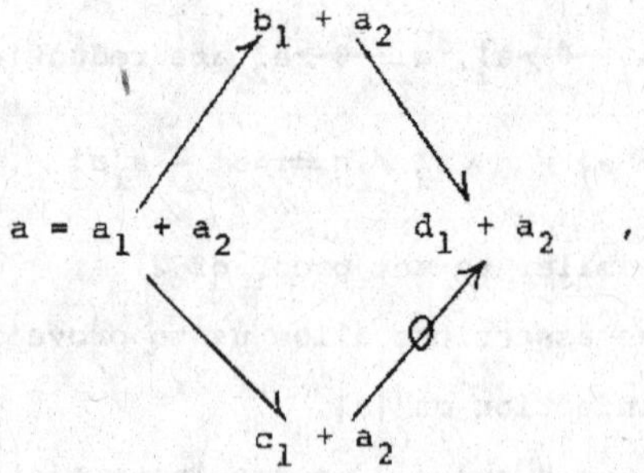

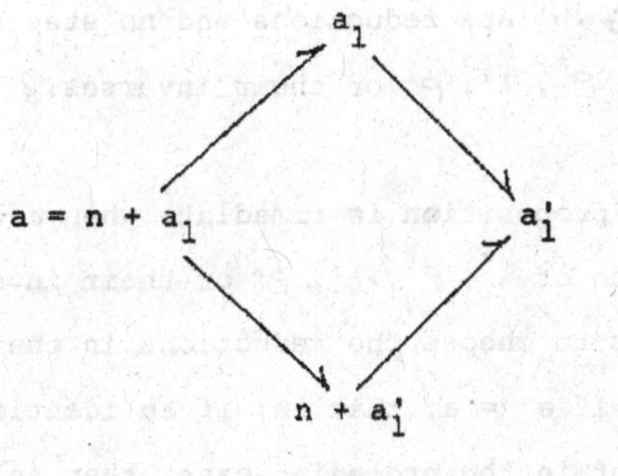

Now we can prove the uniqueness of the value of reduction of a by induction on $|a|$: if $a \longrightarrow b \xrightarrow{\theta} a'$ and $a \longrightarrow c \xrightarrow{\theta} a'$ are two reductions of a we can construct a commutative diagram

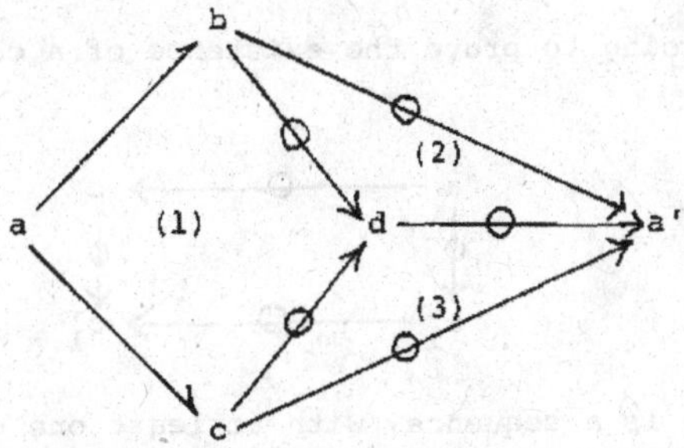

where $b \xrightarrow{\theta} a'$, $d \xrightarrow{\theta} a'$ and $c \xrightarrow{\theta} a'$ are reductions, (1) has been taken commutative following the above result and (2) and (3) are commutative by the induction hypothesis. If a reduction $a \xrightarrow{\theta} a'$ is a sequence of identities, the above argument does not apply, but in this case any reduction is a sequence of identities and the last part of the proposition is trivial.

<u>PROPOSITION 5</u>

Let a and b be elements of $\underline{A}$, $a \longrightarrow b$ an edge of $\underline{T}$ and suppose $\underline{C}$ is coherent. Then, there exists a commutative diagram of type,

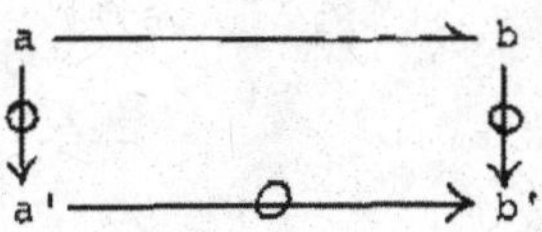

where $a \xrightarrow{\theta} a'$ and $b \xrightarrow{\theta} b'$ are reductions and no step in $a' \xrightarrow{\theta} b'$ is an instantiation of λ^*, ρ^*, λ', ρ' or their inverses.

Proof:

Remark that the proposition is immediate when $a \longrightarrow b$ is an identity or an instantiation of λ^*, ρ^*, λ', ρ' or their inverses because proposition 4 allows us to choose the reductions in the most suitable way for our purposes. If $a' = a$, that is, if an identity is a reduction of a and we are not in the preceding case, then an identity is also a reduction of b and the proposition is immediate.

For the general case we need to prove a preliminary statement: suppose that $a \xrightarrow{\theta} c$ is a path with no instantiation of λ^*, ρ^*, λ', ρ' or their inverses and such that an identity is not a reduction of a. Then, we are going to prove the existence of a commutative diagram of type

$$\begin{array}{ccc} a & \xrightarrow{\quad\theta\quad} & c \\ \theta\big\downarrow & & \big\downarrow\theta \\ a_1' & \xrightarrow{\quad\theta\quad} & c_1' \end{array}$$

where $a \xrightarrow{\theta} a_1'$ is a sequence, with at least one element, of instantiations of λ^*, ρ^*, λ' or ρ', $c \xrightarrow{\theta} c_1'$ is a sequence of identities or instantiations of λ^*, ρ^*, λ' or ρ' and in $a_1' \xrightarrow{\theta} c_1'$ there are no instantiations of λ^*, ρ^*, λ', ρ' or their inverses.

Observe that one consequence of the conditions of the above statement is that $|a_1'| < |a|$ and that if an identity is a reduction of a_1' then any vertex in the path $a_1' \xrightarrow{\theta} c_1'$ has an identity as a reduction (because in it there is no instantiation of λ^*, ρ^*, λ', ρ' or their inverses). This preliminary statement can be proved by induction on $|a|$ following the method outlined in the following diagrams and their analogons:

$$\begin{array}{ccc} a = x + y & \longrightarrow & x' + y = b \\ \Big\downarrow \Theta & & \Big\downarrow \Theta \\ x_1 + y & \longrightarrow & z_1 + y \end{array} ,$$

$$\begin{array}{ccc} a = x + n & \longrightarrow & x' + n = b \\ \Big\downarrow & & \Big\downarrow \\ x & \longrightarrow & x' \end{array} ,$$

$$\begin{array}{ccc} a = x + (y + z) & \longrightarrow & (x + y) + z \\ \Big\downarrow & & \Big\downarrow \\ x' + (y + z) & \longrightarrow & (x' + y) + z \end{array} ,$$

$$\begin{array}{ccc} a = x(y + z) & \longrightarrow & xy + xz \\ \Big\downarrow & & \Big\downarrow \Theta \\ x'(y' + z') & \longrightarrow & x'y' + x'z' \end{array} .$$

From this it is immediate to prove by induction on $|a|$ that if in the path $a \xrightarrow{\Theta} b$ there is no instantiation of λ^*, ρ^*, λ', ρ' or their inverses, then, for $a \xrightarrow{\Theta} a'$ and $b \xrightarrow{\Theta} b'$ reductions there is a commutative diagram of type

$$\begin{array}{ccc} a & \xrightarrow{\Theta} & b \\ \Big\downarrow \Theta & & \Big\downarrow \Theta \\ a' & \xrightarrow{\Theta} & b' \end{array} ,$$

where in $a' \xrightarrow{\Theta} b'$ there is no instantiation of λ^*, ρ^*, λ', ρ' or their inverses. This statement includes all the cases in which the proposition was not proved yet.

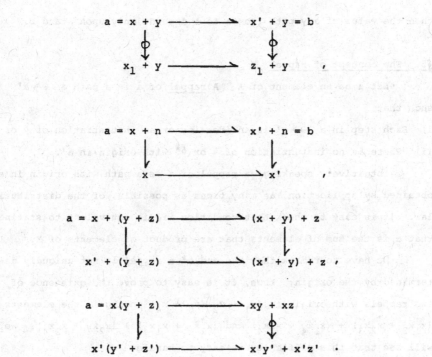

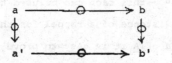

Note that we have not used the hypothesis that the arrows of distributivity are monomorphisms and that an immediate consequence of the above proposition is that if for some element, a of $\underline{A}$, Supp a = n, then the value of any path from a to b depends only upon a and b.

§5. The concept of rappel

Let a be an element of A. A rappel of a is a path $a \xrightarrow{\theta} a'$ such that,

i) Each step in $a \xrightarrow{\theta} a'$ is an identity or an instantiation of δ or $\delta^{\#}$.

ii) There is no instantiation of δ or $\delta^{\#}$ with origin in a'.

Intuitively speaking, a rappel of a is a path with origin in a obtained by application, as many times as possible, of the distributive law. It is easy to check that condition ii) is equivalent to stating that a is the sum of elements that are product of elements of X.

We have to remark that the end of a rappel is not uniquely determined by the origin: thus, it is easy to prove the existence of two rappels with origin in $(x_1 + x_2)(x_3 + x_4)$ ending in the elements $(x_1x_3 + x_2x_3) + (x_1x_4 + x_2x_4)$ and $(x_1x_3 + x_1x_4) + (x_2x_3 + x_2x_4)$. We will see that this difficulty is easy to handle.

In this paragraph we are going to use often induction on siz a-rank a. Note that this number is always non-negative and that any instantiation of α, α', their inverses, γ, and γ' preserves the size and rank and that any instantiation of δ or $\delta^{\#}$ preserves the size and increases the rank, that is, the value of siz a - rank a decreases by instantiations of δ or $\delta^{\#}$: this fact can be used to prove by induction on siz a - rank a the existence of a rappel for the element a.

For any element a of A it is easy to prove that an identity path $a \xrightarrow{\theta} a$ is a rappel if and only if rank a = siz a, and that this is equivalent to stating that a is a sum of products of elements of X.

PROPOSITION 6.

Suppose that $a \longrightarrow b$ is not an instantiation of λ^*, ρ^*, λ', ρ' or their inverses and that $a \longrightarrow c$ is an instantiation of δ or $\delta^{\#}$. Then if C is coherent there exists a commutative diagram of type

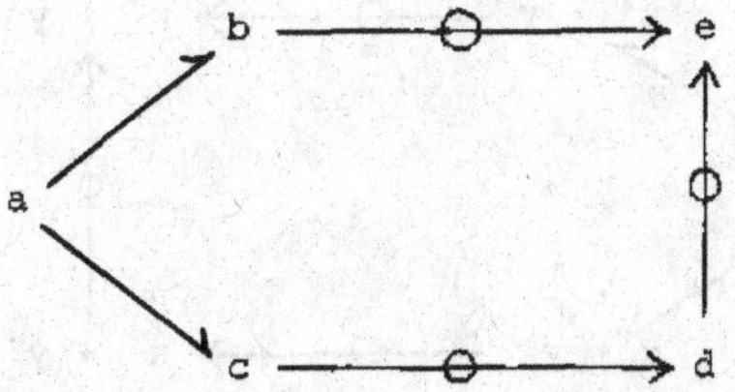

such that $d \xrightarrow{\ominus} e$ is a sequence of identities or instantiations of α, α', λ, ρ, their inverses, γ, and γ', while $b \xrightarrow{\ominus} e$, $c \xrightarrow{\ominus} d$ are sequences of instantiations of δ and $\delta^{\#}$. Moreover in $d \xrightarrow{\ominus} e$ there is some instantiation of λ, ρ, or their inverses if and only if $a \longrightarrow b$ is an instantiation of the same type.

Proof:

The proof can be done by induction on $|a|$ in the form outlined by the following diagrams.

1) In the case

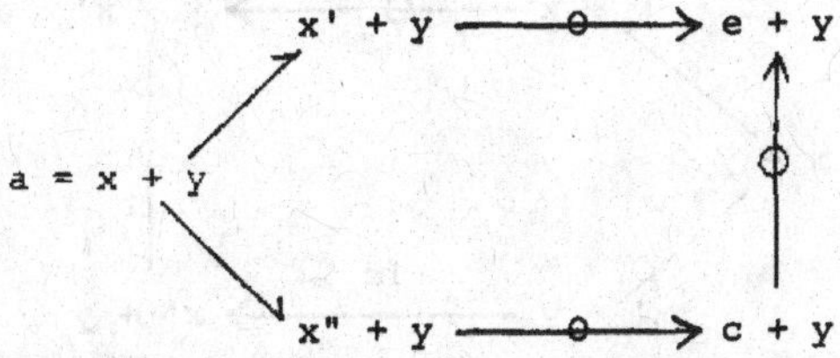

we use the induction hypothesis.

2) In the situation given by

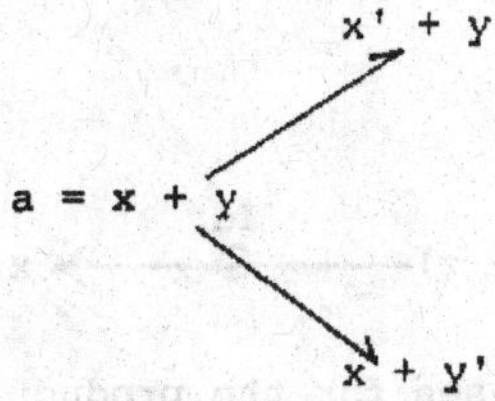

there are two different cases. If $x + y \longrightarrow x' + y$ is an instantiation of δ or $\delta^{\#}$ we can use the construction given by

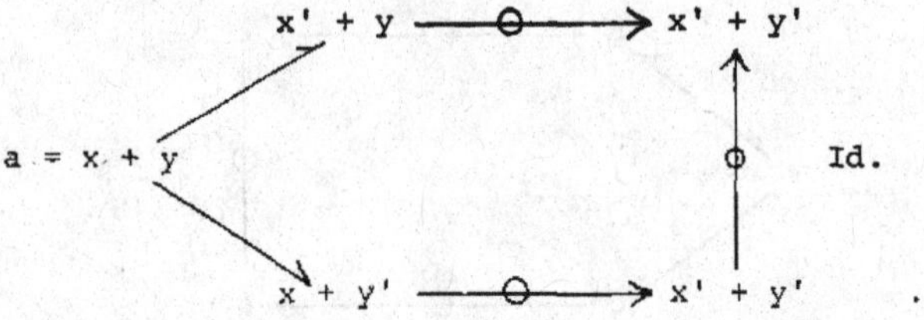

Otherwise, we can take the construction given by

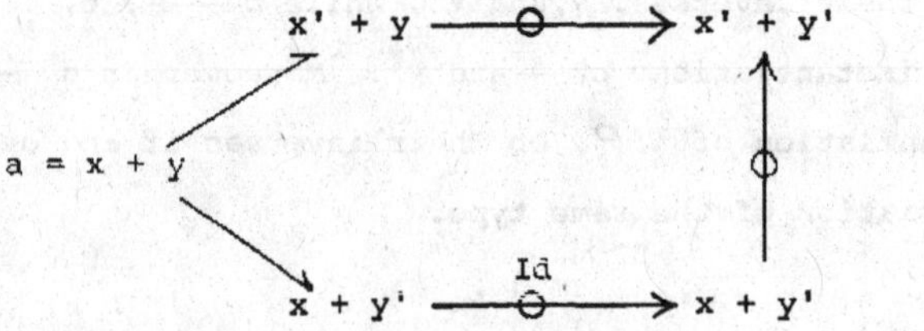

In both constructions we make use of the naturality of ⊕.

3) The naturality of γ allows us to make the construction given by

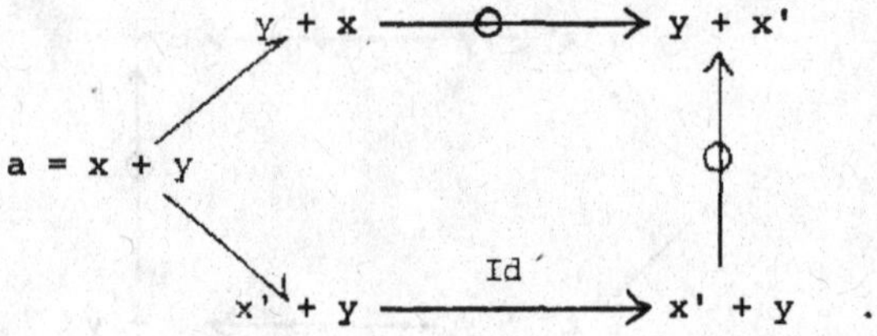

4) The naturality of γ allows the following construction

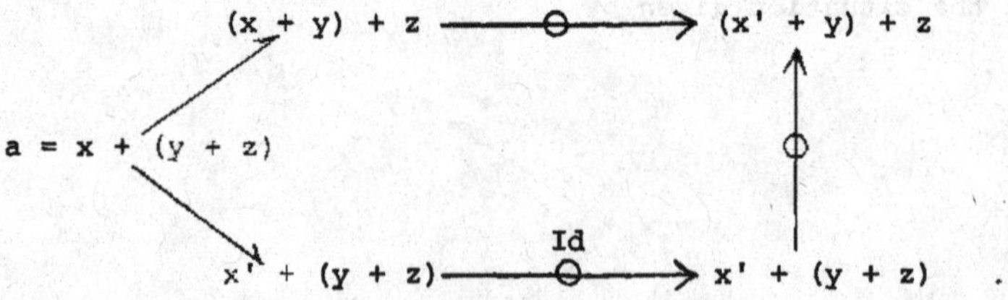

We omit the analogous cases for the product.

5) The commutativity of the diagrams of type (VIII) of the coherence conditions is used in the following construction

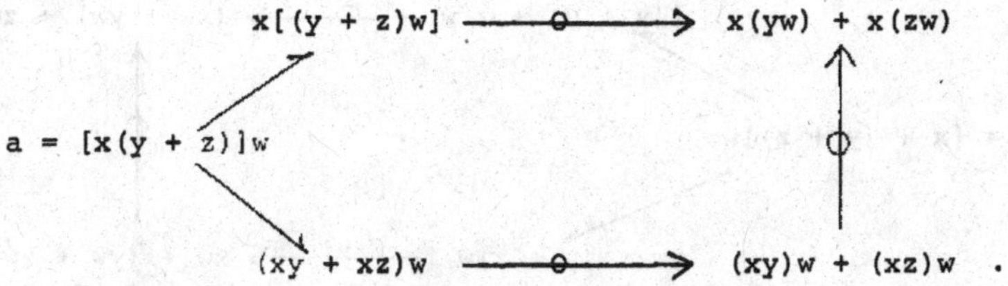

We omit the analogous cases in which we should use the commutativity of (VI) and (VII).

6) The commutativity of the diagrams of type (II) is used in the following construction

We omit the analogous cases in which we should use the commutativity of (I) and (III).

7) The commutativity of the diagrams of type (IX) is used in the following construction

$$(x + y)z + (x + y)w \longrightarrow (xz + yz) + (xw + yw)$$

$$a = (x + y)(z + w)$$

$$x(z + w) + y(z + w) \longrightarrow (xz + xw) + (yz + yw) .$$

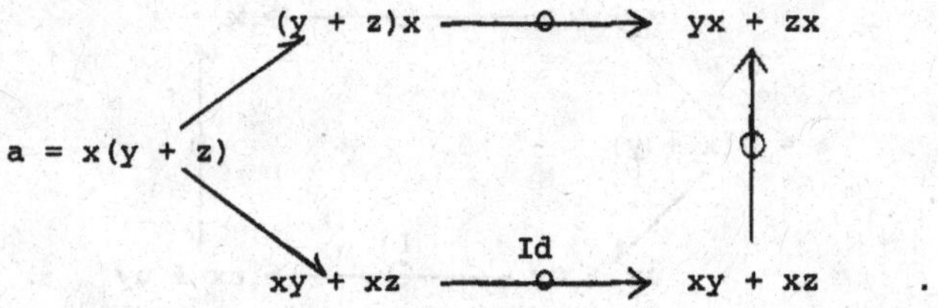

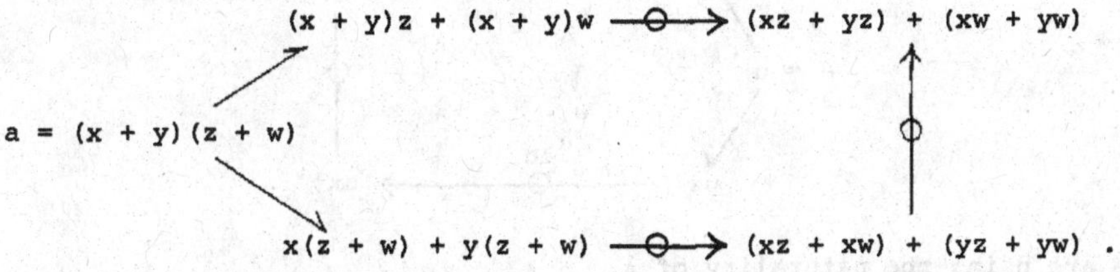

8) The commutativity of the diagrams of type (IV) is used in the following construction

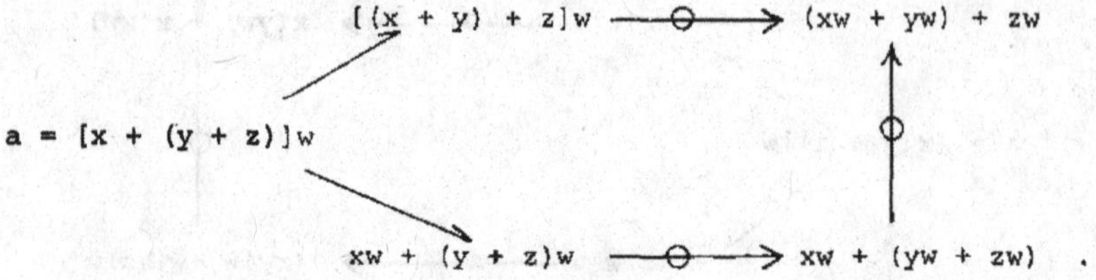

We omit the analogous cases in which we should use the commutativity of (V).

9) We use the commutativity of the diagrams of type (XXIII) in the following construction

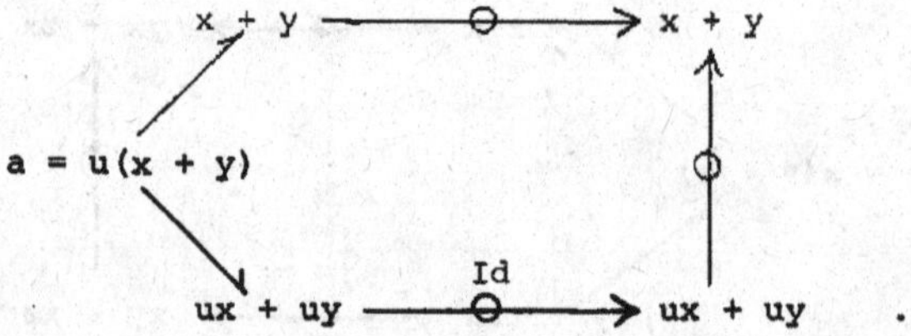

We omit the analogous case in which we should use the commutativity of (XXIV).

10) In the construction

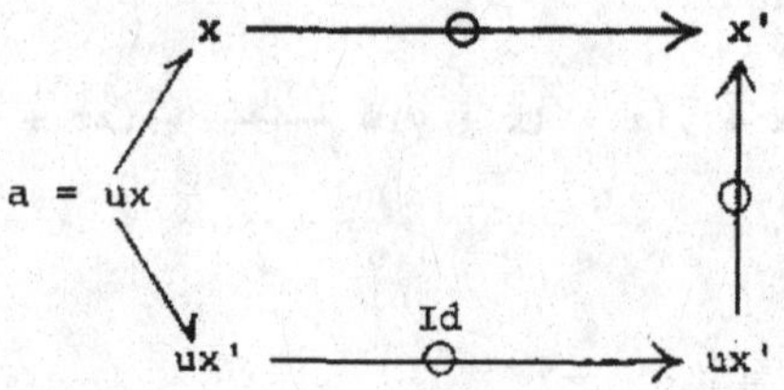

We are using the naturality of λ.

We omit the analogous case in which we should use the naturality of ρ.

11) We use the naturality of δ in the construction given by

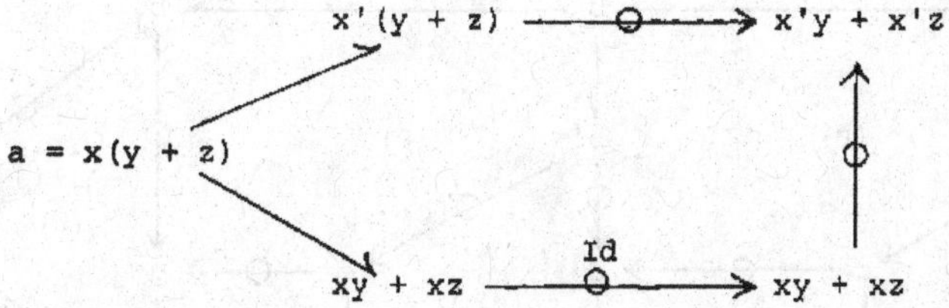

We omit the analogous cases in which we should use the naturality of δ and $\delta^{\#}$.

PROPOSITION 7

Suppose that $\underline{C}$ is coherent, that $a \xrightarrow{\Theta} b$ is a path in whose vertices there is no occurrence of n and that $a \xrightarrow{\Theta} a'$ and $b \xrightarrow{\Theta} b'$ are rappels. Then there exists a commutative diagram of type

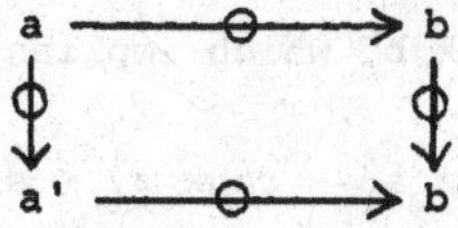

such that $a' \xrightarrow{\Theta} b'$ is a sequence of identities or instantiations of α, α', λ, ρ, their inverses, γ and γ'.

Proof:

The exclusion of n in the vertices of $a \xrightarrow{\Theta} b$ implies that in $a \xrightarrow{\Theta} b$ there is no instantiation of λ^*, ρ^*, λ', ρ' or their inverses and then the form of the proposition allows us to reduce to the case in which $a \xrightarrow{\Theta} b$ is an identity or an instantiation of λ, ρ, α, α', their inverses, γ, γ', δ or $\delta^{\#}$, and this will be done in three parts.

The first part will be proved by induction on siz a - rank a and studies the case in which $a \xrightarrow{\Theta} b$ is an identity or an instantiation of α, α', their inverses, γ and γ', in which case, siz a - rank a = siz b - rank b. The case siz a - rank a = siz b - rank b = 0 is trivial; otherwise we can use the diagram

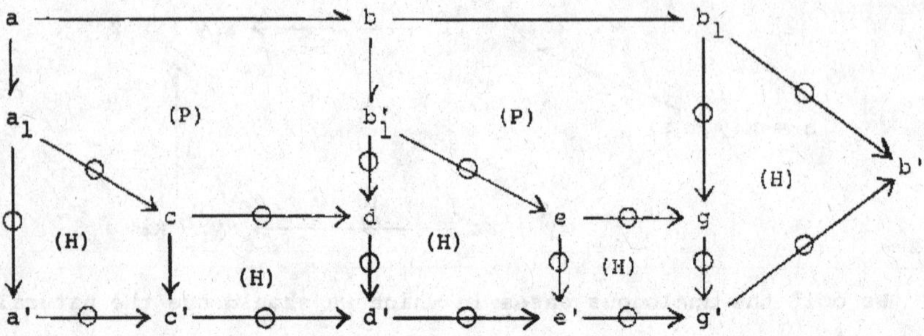

in which a $\longrightarrow a_1 \xrightarrow{\Theta}$ a' and b $\longrightarrow b_1 \xrightarrow{\Theta}$ b' are the given rappels, the diagrams with the symbol (H) have been constructed by the induction hypothesis and the ones with the symbol (P) are constructed using proposition 6. Note that the possibility of the decomposition of the path b $\longrightarrow b_1' \xrightarrow{\Theta}$ d in proposition 6 is assured by the fact that siz a - rank a = siz b - rank b, which implies

$$\text{siz } d - \text{rank } d \leq \text{siz } a_1 - \text{rank } z_1 < \text{siz } a - \text{rank } a.$$

Remark also that to do the induction we have to impose the additional condition that a' $\xrightarrow{\Theta}$ b' is a sequence of identities or instantiations of α, α', their inverses, γ and γ'.

The second part is going to be the proof of the proposition when a $\longrightarrow$ b is an instantiation of λ, ρ, δ or $\delta^{\#}$, and this will be done by induction on siz a - rank a. Remark that for any a of $\underline{A}$,

$$\text{siz(au)} - \text{rank(au)} = (\text{siz } a)(\text{siz } u) - \text{rank } a - \text{rank } u =$$
$$\text{siz } a - \text{rank } a + \text{siz } a - \text{rank } u,$$

where siz a - rank u = 0 if and only if a ϵ X. Hence, in this case, within trivial exception, we can suppose that, siz a - rank a > siz b - rank b, and the proof is outlined in the following diagram

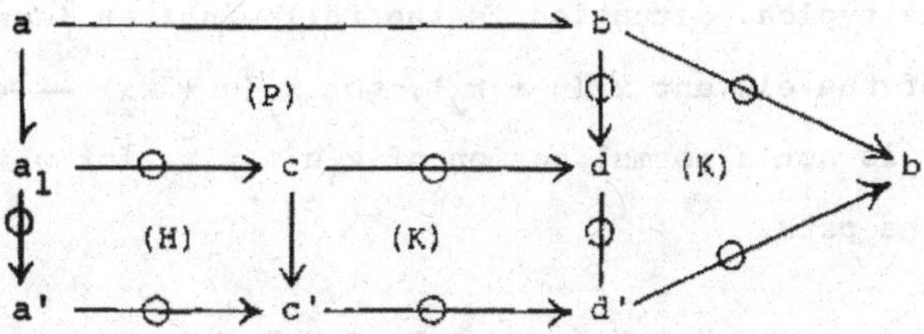

where the symbol (H) in the inside of a diagram means that the induction hypothesis is the reason for the commutativity, (K) the first part of this proposition and the induction hypothesis, and (P) the proposition 6.

The third part is going to be the proof of the proposition for the case in which $a \longrightarrow b$ is an instantiation of λ^{-1} or ρ^{-1} and this is an immediate consequence of the second part and the fact that the bottom path is a sequence of instantiations with formal inverse whose value is an isomorphism.

§6. The concept of normalization

Let a be an element of A. A normalization of a is a path $a \rightarrow\!\!\!\!\!\!\ominus\ a'$ satisfying the following conditions:

i) Any step in $a \rightarrow\!\!\!\!\!\!\ominus\ a'$ is an identity or an instantiation of λ or ρ.

ii) a' is not the origin of any instantiation of λ or ρ.

Intuitively speaking a normalization is a path obtained by application, as many times as possible, of instantiations of λ and ρ. It is easy to prove that if a is the end of a rappel the condition ii) is equivalent to the following: a is the sum of elements that are either u or the product of elements of X different from u. In the general case it is not possible to give a simple characterization of the elements satisfying ii).

The concept of normalization is similar to the concept of reduction or rappel, but it is only useful when applied to elements that are ends of rappels because in this case it eliminates almost completely the occurrences of u in the expression of the elements. In the general case one typical situation is the following: an identity is a normalization of the element $x_1(u + x_2)$, but $x_1(u + x_2) \longrightarrow x_1u + x_1x_2$ and an identity is not a normalization of $x_1u + x_1x_2$ for which a normalization is the path

$$x_1u + x_1x_2 \longrightarrow x_1 + x_1x_2$$

that in fact eliminates all the occurrences of u in the expression of the element.

PROPOSITION 8

Suppose that $\underline{C}$ is coherent and that a is the end of a rappel. Then if $a \xrightarrow{\Theta} a'$ is a normalization, the element a' and the value of $a \xrightarrow{\Theta} a'$ are uniquely determined by a.

Proof:

The proof is similar to (and simpler than) the proof of proposition 4.

PROPOSITION 9

Let a and b be elements of $\underline{A}$ that are ends of rappels, $a \xrightarrow{\Theta} b$ a path whose steps are instantiations of α, α', λ, ρ, their inverses, γ and γ' and $a \xrightarrow{\Theta} a'$, $b \xrightarrow{\Theta} b'$ normalizations of a and b respectively. If $\underline{C}$ is coherent, there exists a commutative diagram of type

$$\begin{array}{ccc} a & \xrightarrow{\Theta} & b \\ \Phi\downarrow & & \downarrow\Phi \\ a' & \xrightarrow{\Theta} & b' \end{array}$$

such that $a' \xrightarrow{\Theta} b'$ is a sequence of identities and instantiations of α, α', their inverses, γ and γ'.

Proof:

It is analogous to (and simpler than) the proof of proposition 5.

Suppose that a is the end of a rappel and that $a \xrightarrow{\Theta} a'$ is a normalization. If a is regular so is a' and if Adec a = $(a_1', \cdots, a_r')$

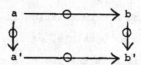

then if $i \neq j$ the set of factors of a_i is different from the set of factors of a_j as is an immediate consequence of the definition of regularity, and, moreover, among the factors of any a_i there is no repetition of elements, as can be also proved almost immediately.

§7. The coherence theorem

We are going to use the results on coherence stated in the Theorem 4.2 of [4], but expressed in a more formal language. We omit a complete proof of the equivalence that is neither difficult nor specially illuminating: in fact it reduces to the same proof given in [4] that holds in the formulation we are going to give. We have to remark that this formulation is different from the one contained in §3 of [2].

Let $\underline{A}'$ be the subset of $\underline{A}$ generated additively by $X - \{n\}$: $\underline{A}'$ is the free $\{+\}$-algebra over $X - \{n\}$. The edges of $\underline{H}$ are a $\{+,.\}$-algebra and hence a $\{+\}$-algebra, and we take as $\underline{H}'$ the subgraph of $\underline{H}$ whose edges are all the elements of the + -subalgebra of the edges of $\underline{H}$ generated by all the elements of the form, $\alpha'_{x,y,z}, \alpha'^{-1}_{x,y,z}, \gamma'_{x,y}$ and 1_x for x,y,z elements of $\underline{A}'$. Suppose that $a \twoheadrightarrow b$ is a path whose vertices are in $\underline{A}'$ and whose steps are elements of $\underline{H}' \cap \underline{T}$, then the Theorem 4.2 of [4] states that if a is regular and $\underline{C}$ is coherent for $\{\alpha', \gamma', \lambda', \rho'\}$ the value of the path $a \twoheadrightarrow b$ only depends upon a and b. Note that an element of $\underline{A}'$ is regular if and only if it is the sum of different elements of X.

We are going to deduce some consequences of that result. Let a be a regular element of $\underline{A}$ in which there is no occurrence of n and suppose that $a \twoheadrightarrow c$ is a path whose steps are identities or instantiations of α', α'^{-1} or γ'. If Adec $a = (a_1, \cdots, a_r)$, the regularity of a implies the relation, $i \neq j \Longrightarrow a \neq a$ and from this and the coherence result above it follows that the value of $a \twoheadrightarrow c$ only depends upon a and c.

Similar consequences hold for the product.

PROPOSITION 10 (Coherence theorem)

If $\underline{C}$ is coherent and a is a regular element of $\underline{A}$, the value of any path a $\ominus\!\!\!\rightarrow$ b depends only upon a and b.

Proof:

Let a $\ominus\!\!\!\rightarrow$ a' and b $\ominus\!\!\!\rightarrow$ b' be reductions. By proposition 5 it is possible to find a commutative diagram of type

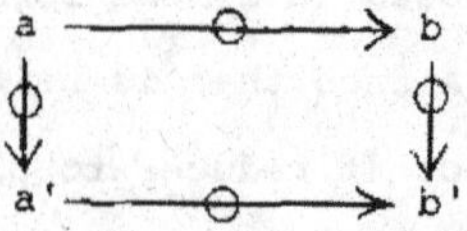

where the value of the columns are isomorphisms that only depend upon a and b, in a' $\ominus\!\!\!\rightarrow$ b' there is no occurrence of instantiations of λ', ρ', λ^*, ρ^* or their inverses, and where all the vertices are n or no n is in the vertices of a' $\ominus\!\!\!\rightarrow$ b'. Hence we are reduced to proving the proposition when in a $\ominus\!\!\!\rightarrow$ b there is no instantiation of λ', ρ', λ^*, ρ^* or their inverses and where the symbol n is not involved in the expression of the vertices: from now on we are going to assume these hypotheses on a $\ominus\!\!\!\rightarrow$ b.

Take now a rappel b $\ominus\!\!\!\rightarrow$ b': the value of it is a monomorphism, hence we are reduced to prove the uniqueness of the value of any path a $\ominus\!\!\!\rightarrow$ b $\ominus\!\!\!\rightarrow$ b', that is, we can (and will) assume the additional hypothesis that b is the end of a rappel. Let a $\ominus\!\!\!\rightarrow$ a' be a rappel: By proposition 7 there is a commutative diagram of type

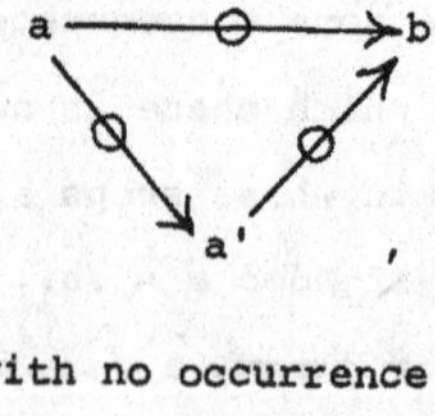

where a' $\ominus\!\!\!\rightarrow$ b is a path with no occurrence of instantiations of δ or $\delta^{\#}$, and we are reduced to prove the uniqueness of the value of a' $\ominus\!\!\!\rightarrow$ b, that is, we are going to assume that a and b are ends of rappels.

Suppose now that a $-\!\ominus\!\!\rightarrow$ a' and b $-\!\ominus\!\!\rightarrow$ b' are normalizations. By propositions 8 and 9 there exists a commutative diagram of type

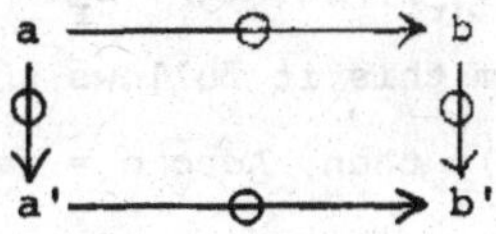

where a', b' and the values of a $-\!\ominus\!\!\rightarrow$ a' and b $-\!\ominus\!\!\rightarrow$ b' depend only upon a and b, and the fact that the values of the columns are isomorphisms allows us to reduce our considerations to the uniqueness of the value of the path a' $-\!\ominus\!\!\rightarrow$ b', that satisfies the conditions indicated in proposition 9. Hence, we are reduced to proving the proposition for the following conditions:

i) Every step in a $-\!\ominus\!\!\rightarrow$ b is an identity or an instantiation of α, α', their inverses, γ and γ'.

ii) Any vertex in the path a $-\!\ominus\!\!\rightarrow$ b is a sum of elements each of which is either or a product of elements of X different from u.

The naturality of $\otimes$ and $\oplus$ implies that any instantiation of α, α^{-1} or γ is commutative with any instantiation of α', α'^{-1} or γ' and this proves the existence of a commutative diagram of type

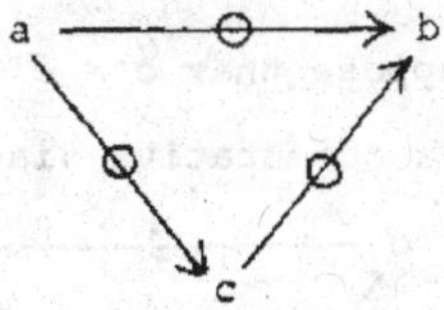

such that in a $-\!\ominus\!\!\rightarrow$ c every step is an identity or an instantiation of α', α'^{-1} or γ', and every step in c $-\!\ominus\!\!\rightarrow$ b is an identity or any instantiation of α, α^{-1} or γ. Our next aim is to prove the uniqueness of c. For this, note the following relations:

1) If d $\longrightarrow$ e is an instantiation of α, α^{-1} or γ, then Apt d = Apt e. This can be proved very easily by induction on $|d|$. Form this it follows that Apt c = Apt b.

2) If $d \longrightarrow e$ is an instantiation of α', α'^{-1} or γ, and d is the end of a rappel, and Adec $d = (d_1, d_2, \cdots, d_r)$, then, Adec $e = (d_{\sigma 1}, d_{\sigma 2}, \cdots, d_{\sigma r})$, with $\sigma \in S_r$. This can be proved by induction on $|d|$. From this it follows that if Adec $a = (a_1, a_2, \cdots, a_r)$, then, Adec $c = (a_{\sigma 1}, \cdots, a_{\sigma r})$ for some $\sigma \in S_r$ and, as we will see later, b determines σ uniquely.

3) If $c \xrightarrow{\ominus} b$ is a sequence of instantiations of α, α^{-1} and γ, c is the end of a rappel, Adec $c = (c_1, \cdots, c_r)$, and Adec $b = (b_1, \cdots, b_r)$, then for $i = 1, 2, \cdots, r$, there is a path $a_i \xrightarrow{\ominus} b_i$ whose steps are identities or instantiations of α, α^{-1} or γ.

From this it follows that for $i = 1, 2, \cdots, r$, $a_{\sigma i} \xrightarrow{\ominus} b_i$, and hence, Supp $a_{\sigma i}$ = Supp b_i. But the regularity of a imposes that, $i \neq j \Longrightarrow$ Supp $a_i \neq$ Supp a_j, and this proves that σi is uniquely determined by the condition, Supp $a_{\sigma i}$ = Supp b_i. Thus b and a determine uniquely Adec c and Apt c and by proposition 1 the element c is uniquely defined.

The uniqueness of the value of $a \xrightarrow{\ominus} c$ has been stated in the remarks of the beginning of §7.

The only thing that remains to be proved is the uniqueness of the value of the path $c \xrightarrow{\ominus} d$ in which all the steps are instantiations of α, α^{-1} and γ. Suppose that $c = c' + c''$, then it is very easy to prove the existence of a commutative diagram of type,

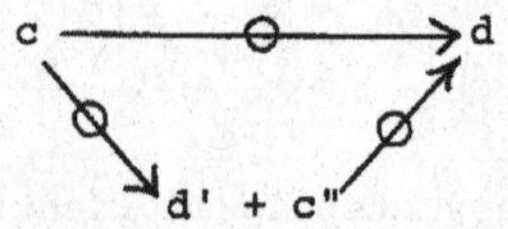

such that in $c \xrightarrow{\ominus} d' + c''$ all the steps are elements of type $\Pi + 1_{c''}$ for some step Π and in $d' + c'' \xrightarrow{\ominus} d$ all the steps are of type $1_{d'} + \Pi$, and with a trivial induction on $|c|$ we are reduced to the case in which c is the product of elements of X, and the proof in this case is analogous to (and easier than) the proof of the uniqueness of the value of path $a \xrightarrow{\ominus} c$.

REFERENCES

[1] G. M. Kelly, "On Mac Lane's conditions for coherence of natural associativities, commutativities, etc.", J. Algebra, 4(1964), 397-402.

[2] G. M. Kelly and S. Mac Lane, "Coherence in closed categories", J. Pure Appl. Algebra, 1(1971), 97-140.

[3] M. Laplaza, "Coherence for categories with associativity, commutativity and distributivity", Bull. Amer. Math. Soc. (to appear).

[4] S. Mac Lane, "Natural associativity and commutativity", Rice Univ. Studies, 49(1963), 28-46.

[5] S. Mac Lane, "Coherence and canonical maps", Symposia Mathematica, IV(1970), 231-241.

MANY-VARIABLE FUNCTORIAL CALCULUS.I.

G.M. Kelly

The University of New South Wales, Kensington 2033, Australia.

Received May 22, 1972

1. Introduction

1.1 The author sees a coherence problem as concerned with an extra structure carried by a category, or more generally by a family of categories; the extra structure consisting in the giving of various functors and natural transformations, subject to various equational axioms. This view may turn out in the end to be too narrow, but the definitive view of "the most general coherence problem" must be at least as wide as this. The present view is in fact extremely wide, and includes all coherence problems known to the author.

In such a structure as we speak of, the basic functors given among the data are functors of many variables and often of mixed variances; as for instance a monoidal structure on A involves functors $\otimes: A^2 \to A$ and $I: A^0 \to A$, while a closed structure also involves a functor $[\ ,\ ]: A^{op} \times A \to A$. The basic natural transformations among the data, such as the associativity $a: (A\otimes B)\otimes C \to A\otimes(B\otimes C)$ in a monoidal structure, connect not these basic functors but others made from them by iterated substitution. Here substitution means the process whereby, from functors

$$T: A \times B^{op} \times C \to D, \quad P: C \times E^{op} \to A, \quad Q: B^{op}\times E \to B, \quad R: F \to C,$$

we get the functor

$$T(P,Q,R): \ C \times E^{op} \times B \times E^{op} \times F \to D$$

whose value at (X,Y,Z,U,V) is $T(P(X,Y), Q(Z,U), R(V))$. Substitution generalizes composition of functors, to which it reduces in the case of functors of one variable. Again, the axioms for the structure, like the pentagonal axiom for a monoidal structure, involve not the basic natural transformations like a but others made from them by "substituting functors into them and them into functors"; thus the pentagonal axiom as given on p. 98 of [7] involves the natural transformations with components

$$a(A\otimes B, C, D):\quad ((A\otimes B)\otimes C)\otimes D \rightarrow (A\otimes B)\otimes(C\otimes D)$$

and

$$a(A,B,C)\otimes D:\quad ((A\otimes B)\otimes C)\otimes D \rightarrow (A\otimes(B\otimes C))\otimes D.$$

An abstract theory of coherence problems needs, therefore, a tidy calculus of substitution for functors of many variables and for suitably general kinds of natural transformations, extending the Godement calculus for functors of a single variable and for ordinary natural transformations. The purpose of this paper is to give such a calculus in the simplest case, when the functors are all covariant and the natural transformations, like the commutativity $c\colon A\otimes B \rightarrow B\otimes A$ in a symmetric monoidal category, do nothing wilder than permuting the variables. Hence the "I" in the title; we hope in later papers to deal with more general kinds of natural transformation. In the last section of this paper we say something about these future plans and the difficulties we have not yet overcome.

1.2 Substitution is not usually seen as a primitive notion; the functor $T(P,Q,R)$ above can be expressed as an ordinary composite $T\circ(P \times Q^{op} \times R)$. In the same way the notion of a complex analytic function $f(z)$ can be expressed in terms of real functions $u(x,y)$ and $v(x,y)$; but one has a calculus of complex functions because they constitute a closed circle of ideas within the larger calculus. So too in functorial calculus there is a closed circle of ideas centred on substitution, with no place for functors like $P \times Q^{op} \times R$ whose codomain is a product $A \times B^{op} \times C$ (or a tensor product, if all our categories are enriched). The point of developing a calculus restricted to substitution-ideas is that these alone suffice for the discussion of coherence problems.

This last assertion, made baldly, is a useful slogan; but does it overstate the case? Since I'm not yet sure, let me now qualify it. Coherence problems can be arranged in a hierarchy according to the degree of generality of the natural transformations they involve. For the lower levels, the assertion is certainly true; moreover at these levels there is a smooth calculus of substitution, which the following paper [5] in this volume reveals as an ideal setting for an abstract discussion of coherence problems. For some higher levels, it remains true that whatever needs to be said can be expressed in terms of substitution alone, but there remain some technical difficulties to be overcome in setting up a smooth calculus of substitution. Finally, for the highest level of generality, the assertion still seems to be true provided we interpret "substitution" fairly liberally; here I can only say, following the analogy of the simpler cases, that I suspect that an appropriate substitution-calculus would provide the right setting. We refer the reader again to the last section of this paper for a glance at the more general cases.

In saying that we can restrict ourselves - with benefit - to substitution, we are in effect claiming that we can get by with a calculus that has no explicit place for functors whose codomain is a product: such as the twisting functor $t: A \times A \to A \times A$ given by $t(A,B) = (B,A)$, or the diagonal functor $\Delta: A \to A{\times}A$ given by $\Delta A = (A,A)$. This must seem at first sight unlikely, since Lawvere's notion of a theory [9], designed to deal with an extra structure carried by a set, makes explicit allowance for the functions analogous to the functors t and Δ. Whereas, however, an algebra may well have a law ab = ba, a structure carried by a category is unlikely to have a <u>functorial</u> law $A\otimes B = B\otimes A$; for this would imply $f\otimes g = g\otimes f\colon A\otimes A \to A\otimes A$, which is not the case in any natural example I know of. Certainly there are respectable cases of monoidal categories where the <u>associativity</u> $(A\otimes B)\otimes C \to A\otimes(B\otimes C)$ is a functorial equality, and also the isomorphisms $I\otimes A \cong A$, $A\otimes I \cong A$; but here we are directly equating two functors $A^3 \to A$, and nothing like t or Δ is involved.

We do of course have things like a <u>natural isomorphism</u> $c_{AB}\colon A\otimes B \to B\otimes A$; this is a natural isomorphism $c\colon \otimes \Rightarrow \otimes t\colon A^2 \to A$, and would seem to involve the functor t. We must therefore explain what we mean by "natural transformation of a general kind".

<u>1.3</u> For $T, S\colon A \to B$ we use a double arrow $f\colon T \Rightarrow S$ to denote a natural transformation in the classical sense, namely a family of morphisms $f(A)\colon TA \to SA$ satisfying the usual naturality condition.

Now consider the commutativity $c_{AB}\colon A\otimes B \to B\otimes A$ in a symmetric monoidal category, the diagonal $d_A\colon A \to A\times A$ in a category with finite products, and the evaluation $e_{AB}\colon [A,B]\otimes A \to B$ in a closed category. Each of these is natural in the sense that, for $f\colon A \to C$ and $g\colon B \to D$, the following diagrams commute:

$$\begin{array}{ccc} A\otimes B & \xrightarrow{c_{AB}} & B\otimes A \\ {\scriptstyle f\otimes g}\downarrow & & \downarrow{\scriptstyle g\otimes f} \\ C\otimes D & \xrightarrow[c_{CD}]{} & D\otimes C \end{array} \qquad\qquad \begin{array}{ccc} A & \xrightarrow{d_A} & A\times A \\ {\scriptstyle f}\downarrow & & \downarrow{\scriptstyle f\times f} \\ C & \xrightarrow[d_C]{} & C\times C \end{array}$$

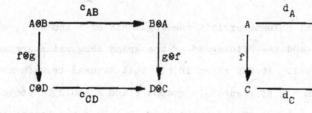

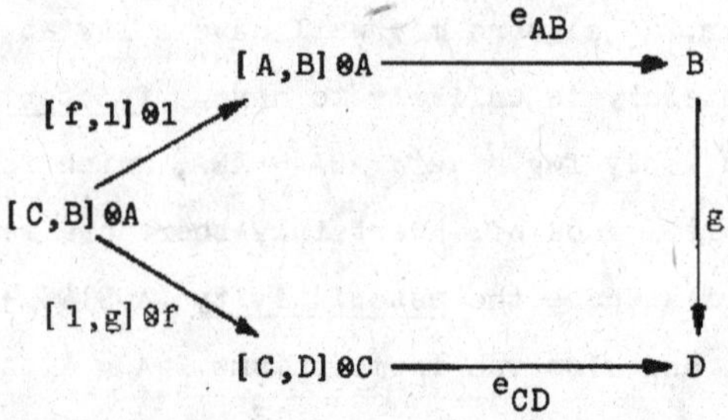

The first two can if we wish be reduced to classical natural transformations $c: \otimes \Rightarrow \otimes t$ and $d: 1 \Rightarrow \times\Delta$ by introducing the functors t and Δ. There is no similar reduction for e_{AB}, where A occurs twice in the domain, once with each variance.

Such natural transformations as e were considered by Eilenberg-Kelly [1]. Writing $T: A^{op} \times A \times A \to A$ for the functor given by $T(A,B,C) = [A,B] \otimes C$ and $1: A \to A$ for the identity functor, they described e as a "natural transformation $e: T \to 1$ of graph $\Gamma e = \varepsilon$ ", where the <u>graph</u> ε of e was the information as to which arguments of T and of 1 were to be set equal in writing the components e_{AB} of e, pictorially ε is representable as

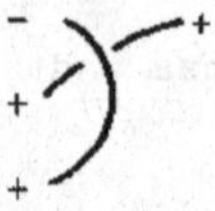

where the vertical columns exhibit the arguments of T and of 1, with their variances, and the "linkages" of the graph show which arguments are to be set equal. It was shown in [1] that natural transformations $f: T \to S$ and $g: S \to R$ of respective graphs ξ and η could be composed to give a composite natural transformation $gf: T \to R$ of the "composite graph" $\eta\xi$, except in the <u>incompatible cases</u> where the composite $\eta\xi$ contained closed loops linking no arguments.

The natural transformations of [1] always had the arguments

linked <u>in pairs</u>, and included as a special case such things as c_{AB}, which in this language would be described as a natural transformation $c\colon \otimes \to \otimes$ of graph $\Gamma c = \gamma$ given by

In such a case as this, where all the functors are covariant, the graph is necessarily a bijection of the arguments of the domain with those of the codomain, and is thus identifiable with a permutation; in particular we can identify the graph γ of c with the non-identity permutation of 2.

The necessity of regarding e, not as a classical natural transformation, but as something autonomous with a graph, suggests doing the same with c. This further suggests doing the same with $d_A\colon A \to A\times A$, calling it a natural transformation $d\colon 1 \to \times$ of graph $\Gamma d = \delta$, where δ is no longer just a permutation but the unique function $2 \to 1$ from the arguments of the codomain to those of the domain. This in turn suggests still further generalization of the notion "natural transformation $f\colon T \to S$ of graph ξ", both in the covariant and the mixed-variance cases, the graph in general being the total information about the arguments set equal; see §4 below. Clearly the concept of "graph" is what allows us to proceed without explicit mention of functors like t and Δ, and opens the way to the kind of calculus we seek. We reiterate that the natural transformations of this paper* are not those of most general graph, but precisely those where the functors are all covariant and the graphs are only permutations. Note that for such graphs incompatibility does not arise.

Graphs were first used in the discussion of coherence

*except in §4

problems in [7]; their success there suggested the treatment in the following paper [5] in this volume, for which the present paper is a preparation. We make two remarks to orient the reader unused to thinking in these terms.

First, the fact that the composite $c_{BA}\ c_{AB}$: $A \otimes B \to B \otimes A \to A \otimes B$ is 1, classically expressed by the equation

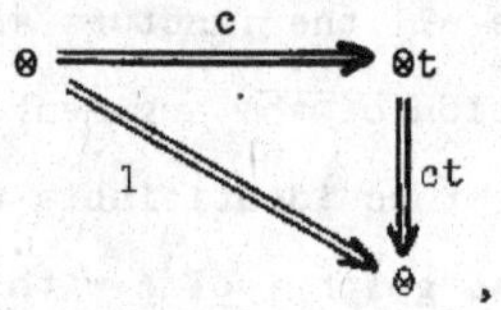

,

is now expressed more simply by the equation

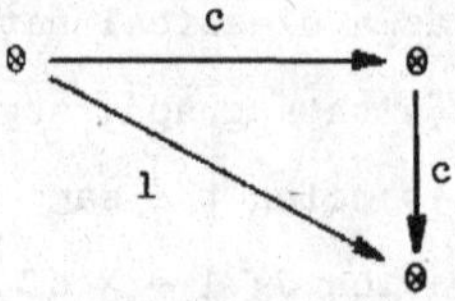

with graph

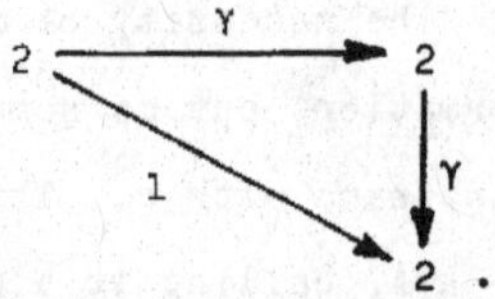

.

Secondly, the reader should note that the "all diagrams commute" result of Mac Lane [12] for symmetric monoidal categories does not mean that $c = 1$: $A \otimes A \to A \otimes A$, which is false for abelian groups; the equality it asserts is that of natural transformations, not of particular components of them. In terms of <u>generic</u> components, it is meaningless to say $c = 1$ because

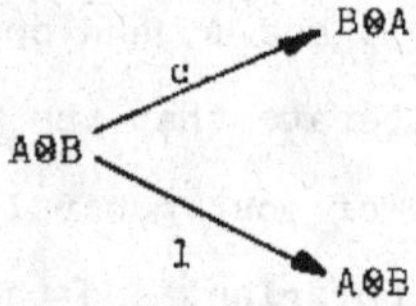

is not a closed diagram. In terms of classical natural transformations, it is meaningless because $c: \otimes \Rightarrow \otimes t$ and $1: \otimes \Rightarrow \otimes$ connect different functors. In our terms, however, it is meaningless because c: $\otimes \to \otimes$ and 1: $\otimes \to \otimes$, while they connect the same functors, have different graphs, with $\Gamma c = \gamma$ and $\Gamma 1 = 1$. For us, natural transformations of different graphs cannot be equated; an "all diagrams commute" result becomes the assertion that the functor Γ, sending each natural transformation to its graph, is faithful.

1.4 We spoke in § 1.1 of a calculus of substitution extending the Godement calculus. The latter calculus is, to use a more recent terminology, the recognition that the category **Cat** of categories (small, or in some universe) has the structure of a 2-category. This, however, says less about **Cat** than the recognition that it is a cartesian closed category, $-\times B$ having the right adjoint $[B,-]$ where $[B,C]$ is the functor category. The 2-category structure follows from the closed structure; for any closed category admits enrichment over itself, so that **Cat** is a **Cat**-category, which is another name for 2-category. In fact vertical composition of natural transformations is embodied in the category $[B,C]$ itself, while both composition of functors and horizontal composition of natural transformations are embodied in the $M\colon [B,C] \times [A,B] \to [A,C]$ arising from the closed structure. The Godement calculus sees only a small part of the structure of **Cat**, dealing only with elements of objects of the special form $[B,C]$.

In the same way, our more general calculus will be expressed in terms of the existence of a closed structure (no longer cartesian, and not even symmetric) on a certain category; the analogue of the M above will then provide our "calculus of substitution", but it is the whole closed structure that we need for our abstract discussion of coherence problems.

2. The single-category case

2.1 We begin, purely for pedagogic simplicity, with the calculus needed to describe extra structures carried by a single category (with all functors moreover covariant and all graphs merely permutations). If the structure is to be carried by A, this will mean having functors of the form $A^n \to A$, where $A^n = A \times \ldots \times A$, while A^0 = the unit category I with one object and one morphism. It turns out however, that the right thing to consider is rather functors $A^n \to B$. Note that a functor $A^0 \to B$ is just an object of B, and that a natural transformation between two such functors is just a morphism in B. We use this section to sketch what is to come, with motivation, before setting it down formally.

$\underline{\underline{N}}$ is the set of natural numbers, including 0. We ambiguously denote by n the natural number n or the finite set {1,2, ..., n}. Write $\underline{\underline{P}}$ for the category with $\underline{\underline{N}}$ as set of objects, with no morphisms $n \to m$ for $n \neq m$, and with the permutations of n as the morphisms $n \to n$.

Given categories B and C we form a "generalized functor category" $\{B,C\}$. An object is an $n \in \underline{\underline{N}}$ together with a functor $T: B^n \to C$; we call n the type of T. There are no morphisms $(n,T) \to (m,S)$ unless $n = m$, that is, unless T and S are of the same type. When they are, a morphism $(n,T) \to (n,S)$ is a permutation ξ of n together with a natural transformation $f: T \to S$ of graph ξ. Such an f has components

$$(2.1) \qquad f(A_1, \ldots, A_n): T(A_{\xi 1}, \ldots, A_{\xi n}) \to S(A_1, \ldots, A_n).$$

Setting $\Gamma(n,T) = n$ and $\Gamma(\xi,f) = \xi$ we get a functor $\Gamma: \{B,C\} \to \underline{\underline{P}}$, exhibiting $\{B,C\}$ as an object of the category $\underline{\underline{Cat}}/\underline{\underline{P}}$ of categories over $\underline{\underline{P}}$, with augmentation Γ.

The assignment $B,C \mapsto \{B,C\}$ is easily seen to provide a functor $\underline{\underline{Cat}}^{op} \times \underline{\underline{Cat}} \to \underline{\underline{Cat}}/\underline{\underline{P}}$, whose evident continuity in C ensures that it will have a left adjoint: that is, there is a functor

$$(2.2) \qquad \circ : \underline{\underline{\mathrm{Cat}}}/\underline{\underline{\mathrm{P}}} \times \underline{\underline{\mathrm{Cat}}} \to \underline{\underline{\mathrm{Cat}}}$$

and a natural isomorphism

$$(2.3) \qquad \underline{\underline{\mathrm{Cat}}}\,(\mathcal{A}\circ\mathcal{B},\mathcal{C}) \cong \underline{\underline{\mathrm{Cat}}}/\underline{\underline{\mathrm{P}}}\,(\mathcal{A}, \{\mathcal{B},\mathcal{C}\}).$$

It is in fact easy to see explicitly what $\mathcal{A}\circ\mathcal{B}$ must be. Write Γ for all augmentations over $\underline{\underline{\mathrm{P}}}$. To give an element Φ of the right side of (2.3) one must give a functor $\Phi A\colon \mathcal{B}^n \to \mathcal{C}$ for each $A \in \mathcal{A}$ with $\Gamma A = n$, and a natural transformation $\Phi f\colon \Phi A \to \Phi A'$ of graph ξ for each $f\colon A \to A'$ with $\Gamma f = \xi$. To give ΦA one must give objects $\Phi A(B_1, \ldots, B_n)$ of $\mathcal{C}$ and morphisms $\Phi A(g_1, \ldots, g_n)\colon \Phi A(B_1, \ldots, B_n) \to \Phi A(B_1', \ldots, B_n')$, where $g_i\colon B_i \to B_i'$ in $\mathcal{B}$. To give Φf one must give its components, that is, morphisms $\Phi f(B_1, \ldots, B_n)\colon \Phi A(B_{\xi 1}, \ldots, B_{\xi n}) \to \Phi A'(B_1, \ldots, B_n)$. These data are to satisfy the conditions making ΦA functorial, making Φf natural, making $\Phi(f'f)$ equal to $(\Phi f')(\Phi f)$, and making $\Phi(1_A)$ equal to $1_{\Phi A}$.

This means that $\mathcal{A}\circ\mathcal{B}$ is to have objects of the form $A[B_1, \ldots, B_n]$ where $A \in \mathcal{A}$, $B_i \in \mathcal{B}$, and $\Gamma A = n$; its morphisms are to be generated by $A[g_1, \ldots, g_n]\colon A[B_1, \ldots, B_n] \to A[B_1', \ldots, B_n']$ and $f[B_1, \ldots, B_n]\colon A[B_{\xi 1}, \ldots, B_{\xi n}] \to A'[B_1, \ldots, B_n]$; these generators are to satisfy relations corresponding to the conditions at the end of the last paragraph. One of these, the naturality of Φf, gives the relation

$$(2.4) \qquad \begin{array}{ccc} A[B_{\xi 1}, \ldots, B_{\xi n}] & \xrightarrow{\;f[B_1, \ldots, B_n]\;} & A'[B_1, \ldots, B_n] \\ {\scriptstyle A[g_{\xi 1}, \ldots, g_{\xi n}]}\downarrow & & \downarrow{\scriptstyle A'[g_1,\ldots,g_n]} \\ A[B'_{\xi 1}, \ldots, B'_{\xi n}] & \xrightarrow[\;f[B_1', \ldots, B_n']\;]{} & A'[B_1', \ldots, B_n'] \end{array}$$

Writing $f[g_1, \ldots, g_n]$ for the diagonal of this, one easily verifies that this is the most general morphism of $\mathcal{A}\circ\mathcal{B}$, and that the edges of

(2.4) are just the special cases in which, in accordance with the usual convention, A and B_1 stand for 1_A and 1_{B_1}.

Next, regard $\underline{\underline{Cat}}$ as a full subcategory of $\underline{\underline{Cat}}/\underline{\underline{P}}$ by giving to $A \in \underline{\underline{Cat}}$ the augmentation $A \to \underline{\underline{P}}$ consisting of the constant functor at $0 \in \underline{\underline{P}}$. Then (2.2) admits an immediate extension to a functor

(2.5) $\qquad \circ\colon \underline{\underline{Cat}}/\underline{\underline{P}} \times \underline{\underline{Cat}}/\underline{\underline{P}} \to \underline{\underline{Cat}}/\underline{\underline{P}};$

one has only to define $A \circ B$ as before, ignoring the augmentation of B, and then augment $A \circ B$ by setting $\Gamma(A[B_1, \ldots, B_n]) = \Gamma B_1 + \ldots + \Gamma B_n$ on objects, with an appropriate definition on morphisms. One easily verifies that the bifunctor $\circ$ on $\underline{\underline{Cat}}/\underline{\underline{P}}$ is coherently associative with a coherent identity J, making $\underline{\underline{Cat}}/\underline{\underline{P}}$ into a monoidal category - highly unsymmetric, of course. Finally one verifies that our original functor

(2.6) $\qquad \{\ ,\ \}\colon \underline{\underline{Cat}}^{op} \times \underline{\underline{Cat}} \to \underline{\underline{Cat}}/\underline{\underline{P}}$

itself extends to

(2.7) $\qquad \{\ ,\ \}\colon (\underline{\underline{Cat}}/\underline{\underline{P}})^{op} \times (\underline{\underline{Cat}}/\underline{\underline{P}}) \to \underline{\underline{Cat}}/\underline{\underline{P}},$

and the natural isomorphism (2.3) to

(2.8) $\qquad \underline{\underline{Cat}}/\underline{\underline{P}}\,(A \circ B, C) \cong \underline{\underline{Cat}}/\underline{\underline{P}}(A, \{B, C\});$

exhibiting $\underline{\underline{Cat}}/\underline{\underline{P}}$ as a closed category. This is our goal; the substitution-calculus as such is embodied in the functor $\mu\colon \{B,C\} \circ \{A,B\} \to \{A,C\}$ arising from this closed structure. It is useful for later purposes to observe that $\circ$ and $\{\ ,\ \}$ are actually 2-functors, and that (2.8) is a 2-natural isomorphism of categories; that is, a 2-adjunction (in the older sense of $\underline{\underline{Cat}}$-adjunction, not that of Gray [4]).

We now set this down formally.

2.2 We need some notation for the process of substituting permutations in permutations. For $n \in \underline{\underline{N}}$, define a functor $Y_n: \underline{\underline{P}}^n \to \underline{\underline{P}}$ by

$$Y_n(m_1, \ldots, m_n) = m_1 + \ldots + m_n,$$

(2.9)

$$Y_n(\eta_1, \ldots, \eta_n) = \eta_1 + \ldots + \eta_n;$$

here η_i is a permutation of m_i and $\eta_1 + \ldots + \eta_n$ is the permutation of $m_1 + \ldots + m_n$ which effects η_1 on the first block of m_1 elements, η_2 on the next block of m_2, and so on. For a permutation ξ of n, define a natural transformation $Y_\xi: Y_n \to Y_n$ of graph ξ whose component

(2.10) $$Y_\xi(m_1, \ldots, m_n): m_{\xi 1} + \ldots + m_{\xi n} \to m_1 + \ldots + m_n$$

is the evident permutation that "permutes the blocks according to ξ".

We introduce the above notation because we have occasion below to refer explicitly, although briefly, to Y_n and Y_ξ. For a working notation, however, we drop the Y and set

(2.11) $$n(m_1, \ldots, m_n) = Y_n(m_1, \ldots, m_n) = m_1 + \ldots + m_n,$$

(2.12) $$n(\eta_1, \ldots, \eta_n) = Y_n(\eta_1, \ldots, \eta_n) = \eta_1 + \ldots + \eta_n,$$

(2.13) $$\xi(m_1, \ldots, m_n) = Y_\xi(m_1, \ldots, m_n).$$

The naturality of Y_ξ is expressed by the commutativity of

(2.14) $$\begin{array}{ccc} n(m_{\xi 1}, \ldots, m_{\xi n}) & \xrightarrow{\xi(m_1, \ldots, m_n)} & n(m_1, \ldots, m_n) \\ {\scriptstyle n(\eta_{\xi 1}, \ldots, \eta_{\xi n})}\downarrow & & \downarrow{\scriptstyle n(\eta_1, \ldots, \eta_n)} \\ n(m_{\xi 1}, \ldots, m_{\xi n}) & \xrightarrow[\xi(m_1, \ldots, m_n)]{} & n(m_1, \ldots, m_n) \end{array}$$

We denote the diagonal of this by

(2.15) $$\xi(\eta_1, \ldots, \eta_n);$$

so (2.15) is the permutation of $m_1 + \ldots + m_n$ which permutes the blocks according to ξ and at the same time effects the permutation η_i on the appropriate block. Observe that (2.12) is just the special case of (2.15) when ξ is replaced by 1_n and that (2.13) is the special case of (2.15) when η_i is replaced by 1_{m_i}. We repeat for the last time that all our notation is consistent with the convention that the name of an object is also the name of its identity morphism.

We leave the reader to verify such properties of the expression (2.15) as are needed in the sequel; to list them would be a waste of space since they are entirely summed up in the assertion that $\xi[\eta_1, \ldots, \eta_n] \mapsto \xi(\eta_1, \ldots, \eta_n)$ is a functor $\mu: \underline{\underline{P}} \circ \underline{\underline{P}} \to \underline{\underline{P}}$ making $\underline{\underline{P}}$ a $\circ$-monoid in $\underline{\underline{Cat}}/\underline{\underline{P}}$ with identity 1_1 (the identity permutation of 1).

<u>2.3</u> We give the formal definition of $A \circ B$ for $A, B \in \underline{\underline{Cat}}/\underline{\underline{P}}$; all augmentations over $\underline{\underline{P}}$ are denoted by Γ.

As a category, $A \circ B$ has objects $A[B_1, \ldots, B_n]$ where $A \in A$ with $\Gamma A = n$ and $B_i \in B$. Note that this includes objects $A[\ ]$ where $\Gamma A = 0$. There are no morphisms $A[B_1, \ldots, B_n] \to A'[B'_1, \ldots, B'_m]$ for $m \neq n$. A morphism $A[B_1, \ldots, B_n] \to A'[B'_1, \ldots, B'_n]$, where $\Gamma A = \Gamma A' = n$, consists of a morphism $f: A \to A'$ in A with $\Gamma f = \xi$ say, together with morphisms $g_i: B_{\xi^{-1}i} \to B'_i$ in B; this morphism is denoted by

(2.16) $\quad f[g_1, \ldots, g_n]: A[B_1, \ldots, B_n] \to A'[B'_1, \ldots, B'_n]$.

(It would look more like (2.4) if we wrote

$$f[g_1, \ldots, g_n]: A[B_{\xi 1}, \ldots B_{\xi n}] \to A'[B'_1, \ldots, B'_n]$$

where $f: A \to A'$ with $\Gamma f = \xi$ and where $g_i: B_i \to B'_i$.) The composite of (2.16) with $h[k_1, \ldots, k_n]: A'[B'_1, \ldots, B'_n] \to A''[B''_1, \ldots, B''_n]$, where $\Gamma h = \zeta$, is $(hf)[p_1, \ldots, p_n]$ where p_i is the composite

$$(2.17) \qquad B_{\xi^{-1}\zeta^{-1}i} \xrightarrow{\;g_{\zeta^{-1}i}\;} B'_{\zeta^{-1}i} \xrightarrow{\;k_i\;} B''_i .$$

Clearly $A \circ B$ is a category, with identities $A[B_1, \ldots, B_n] = 1_A[1_{B_1}, \ldots, 1_{B_n}]$; note that $\Gamma 1_A = 1_n$ since Γ is a functor.

We make $A \circ B$ into a category over **P** by setting

$$(2.18) \qquad \Gamma(A[B_1, \ldots, B_n]) = \Gamma A(\Gamma B_1, \ldots, \Gamma B_n) = \Gamma B_1 + \ldots + \Gamma B_n,$$

$$(2.19) \qquad \Gamma(f[g_1, \ldots, g_n]) = \Gamma f(\Gamma g_1, \ldots, \Gamma g_n),$$

in the notation of §2.2.

The operation $\circ$ becomes a functor **Cat**/**P** × **Cat**/**P** → **Cat**/**P** when we define, for functors $T\colon A \to A'$ and $S\colon B \to B'$ over **P**, the functor $T \circ S\colon A \circ B \to A' \circ B'$ by

$$(2.20) \qquad \begin{aligned} T \circ S\,(A[B_1, \ldots, B_n]) &= TA[SB_1, \ldots, SB_n], \\ T \circ S\,(f[g_1, \ldots, g_n]) &= Tf[Sg_1, \ldots, Sg_n]. \end{aligned}$$

Cat/**P** is actually a 2-category; a 2-cell is a natural transformation $\alpha\colon T \Rightarrow T'\colon A \to A'$ <u>over **P**</u>, that is, one for which we have commutativity in

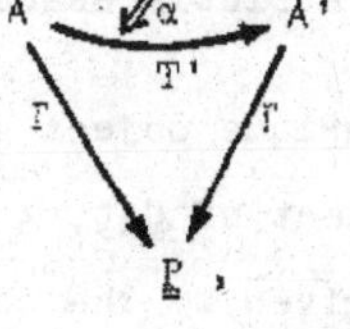

,

which means that $\Gamma(\alpha A) = 1_{\Gamma A}$. The functor $\circ$ becomes a 2-functor when, for $\alpha\colon T \Rightarrow T'$ and $\beta\colon S \Rightarrow S'$ over **P**, we set

$$(2.21) \qquad \alpha \circ \beta\,(A[B_1, \ldots, B_n]) = \alpha A[\beta B_1, \ldots, \beta B_n].$$

Identifying **Cat** as in §2.1 with the full subcategory of **Cat**/**P** consisting of those A whose augmentation is the constant functor at 0, it is immediate from (2.18) that $\bullet$ restricts to a 2-functor $\circ: \underline{\underline{Cat}}/\underline{P} \times \underline{\underline{Cat}} \to \underline{\underline{Cat}}$.

Another way of looking at $A \circ B$ is instructive. For $B \in \underline{\underline{Cat}}$, form a category over $\underline{P}$ called MB ("multi-B"); an object is an n together with an n-ad $[B_1, \ldots, B_n]$; a morphism $[B_1, \ldots, B_n] \to [B'_1, \ldots, B'_n]$ is an ξ together with an n-ad $[g_1, \ldots, g_n]$ where $g_i : B_{\xi^{-1}i} \to B'_i$. Then $A \circ B$ is just the fibred product over $\underline{P}$ of A and MB; when B itself is over $\underline{P}$, MB gets a second augmentation over $\underline{P}$ used to give the augmentation of $A \circ B$. Note that as a category MB is just $\underline{P} \circ B$.

There is a 2-natural isomorphism $(A \circ B) \circ C \cong A \circ (B \circ C)$, sending $(A[B_1, \ldots, B_n])[C_1, \ldots, C_m]$ to

$$A[B_1[C_1, \ldots, C_{m_1}], B_2[C_{m_1+1}, \ldots, C_{m_1+m_2}], \ldots, B_n[\ldots, C_m]];$$

here $\Gamma B_i = m_i$ and $m_1 + \ldots + m_n = m$. The isomorphism is defined similarly on morphisms, and its 2-naturality is immediate. It clearly satisfies the pentagonal condition for coherence; we could in fact define $A \circ B \circ C$ without parentheses to have objects of the form $[A][B_1, \ldots, B_n][C_1, \ldots, C_m]$, and so on. We shall normally suppress the isomorphism and treat $\circ$ as strictly associative where convenient.

Denote by 1 both the unique object and the unique morphism of the unit category I. As an object of $\underline{\underline{Cat}}$, I is also an object of $\underline{\underline{Cat}}/\underline{P}$ with augmentation $I \to \underline{P}$ given by the constant functor at 0. Define a new object J of $\underline{\underline{Cat}}/\underline{P}$ to be the category I but with a different augmentation $I \to \underline{P}$, namely the constant functor at $1 \in \underline{P}$. Then J is a coherent two-sided identity for $\circ$; the isomorphism $J \circ A \to A$ sends $1[A]$ to A and $1[f]$ to f, while the isomorphism $A \circ J \to A$ sends $A[1,1, \ldots, 1]$ to A and $f[1,1, \ldots, 1]$ to f.

Thus we have exhibited $\underline{\underline{Cat}}/\underline{\underline{P}}$ as a monoidal category - or rather monoidal 2-category - with "tensor product" $\circ$. It is by no means symmetric; indeed when $A \in \underline{\underline{Cat}}$ we have $A \circ B \cong A$, the isomorphism sending $A[\ ]$ to A. Note that, as this example shows, the functor $A\circ-$ fails to preserve colimits and therefore has no right adjoint; yet as we promised in (2.8) $-\circ B$ has a right adjoint $\{B,-\}$.

2.4 We now proceed to the formal definition of $\{B,C\}$.

Observe that any function ϕ: $n \to m$ induces a functor $B^{\phi}: B^m \to B^n$, namely $(B_1, \ldots, B_m) \mapsto (B_{\phi 1}, \ldots, B_{\phi n})$; we have $B^{\phi\psi} = B^{\psi}B^{\phi}$ and $B^{id} = id$. We are concerned here only with the case where $m = n$ and ϕ is a permutation.

For T,S: $B^n \to C$ and for a permutation ξ of n, a natural transformation f: $T \to S$ of graph ξ means a classical natural transformation f: $TB^{\xi} \Rightarrow S$, which can be pictured as a 2-cell

(2.22)

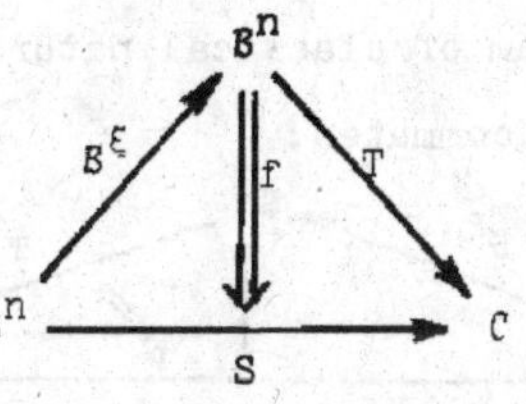

Equivalently, f is given by components

(2.23) $f(B_1, \ldots, B_n): T(B_{\xi 1}, \ldots, B_{\xi n}) \to S(B_1, \ldots, B_n)$,

natural in the usual sense in each B_i. If also R: $B^n \to C$ and if g: $S \to R$ is a natural transformation of graph η, we define the composite gf: $T \to R$ of graph $\eta\xi$ to be the classical composite

$$TB^{\eta\xi} = TB^{\xi}B^{\eta} \xrightarrow[fB^{\eta}]{} SB^{\eta} \xrightarrow[g]{} R;$$

its component $(gf)(B_1, \ldots, B_n)$ is therefore the composite

$$(2.24)\; T(B_{\eta\xi 1},\ldots,B_{\eta\xi n}) \xrightarrow[f(B_{\eta 1},\ldots,B_{\eta n})]{} S(B_{\eta 1},\ldots,B_{\eta n}) \xrightarrow[g(B_1,\ldots,B_n)]{} R(B_1,\ldots,B_n).$$

Clearly this composition is associative, and there is an identity natural transformation $1: T \to T$ with identity graph and identity components.

Given now $\mathcal{B},\mathcal{C} \in \underline{\underline{Cat}}/\underline{\underline{P}}$ we define the category $\{\mathcal{B},\mathcal{C}\}$. An object is an $n \in \underline{\underline{N}}$ together with a functor $T: \mathcal{B}^n \to \mathcal{C}$ making commutative the diagram

(2.25)

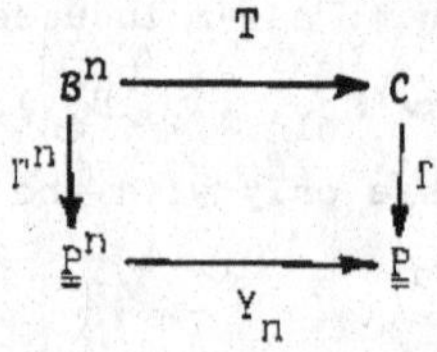

where Y_n is as defined in §2.2. There are no morphisms $(n,T) \to (m,S)$ for $n \neq m$. A morphism $(n,T) \to (n,S)$ consists of a permutation ξ of n together with a natural transformation $f: T \to S$ of graph ξ, such that the following diagram of classical natural transformations (represented by 2-cells) commutes:

(2.26)

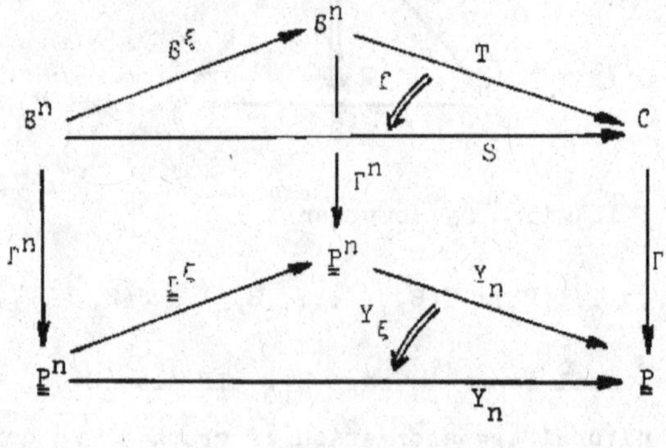

Composition in $\{\mathcal{B},\mathcal{C}\}$ is the composition of natural transformations described in the last paragraph; (2.26) for the composite is automatic when it holds for the factors. Finally $\{\mathcal{B},\mathcal{C}\}$ is made into a category over $\underline{\underline{P}}$ by giving it the augmentation $\Gamma(n,T) = n$, $\Gamma(\xi,f) = \xi$. We shall usually abbreviate (n,T) to T and (ξ,f) to f.

Using the definitions (2.11) - (2.13) we may write (2.25) and (2.26) in terms of components; (2.25) reduces to the first two assertions below and (2.26) to the third:

(2.27) $\qquad \Gamma(T(B_1, \ldots, B_n)) = \Gamma T(\Gamma B_1, \ldots, \Gamma B_n),$

(2.28) $\qquad \Gamma(T(g_1, \ldots, g_n)) = \Gamma T(\Gamma g_1, \ldots, \Gamma g_n),$

(2.29) $\qquad \Gamma(f(B_1, \ldots, B_n)) = \Gamma f(\Gamma B_1, \ldots, \Gamma B_n).$

When $\mathcal{B}, \mathcal{C} \in \underline{\underline{Cat}}$ these are automatically and trivially fulfilled, so that $\{\mathcal{B},\mathcal{C}\}$ is just as given in §2.1; it has as objects <u>all</u> the functors $T\colon \mathcal{B}^n \to \mathcal{C}$ of all types n, and as morphisms <u>all</u> the natural transformations $f\colon T \to S$ of all graphs ξ.

$\{\ ,\ \}$ becomes a functor $(\underline{\underline{Cat}}/\underline{\underline{P}})^{op} \times (\underline{\underline{Cat}}/\underline{\underline{P}}) \to \underline{\underline{Cat}}/\underline{\underline{P}}$ when, for functors $U\colon \mathcal{D} \to \mathcal{B}$ and $V\colon \mathcal{C} \to \mathcal{E}$ over $\underline{\underline{P}}$, we define $\{U,V\}\colon \{\mathcal{B},\mathcal{C}\} \to \{\mathcal{D},\mathcal{E}\}$ as the functor over $\underline{\underline{P}}$ sending T to VTU^n and sending f to the 2-cell

(2.30)

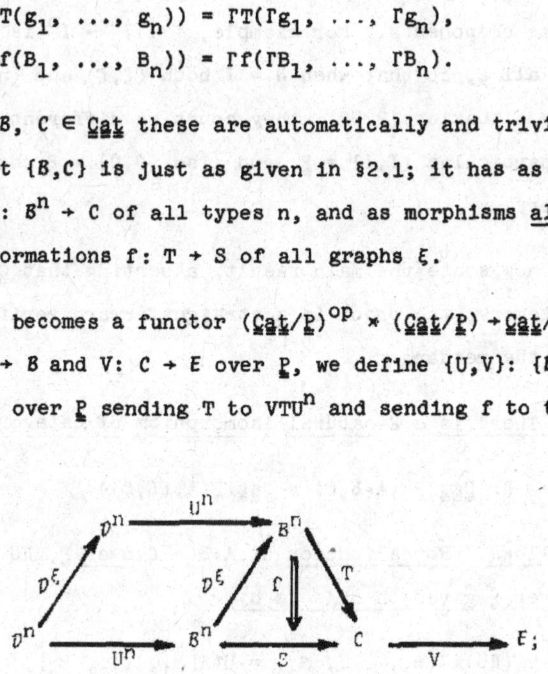

in other words,

(2.31) $\qquad (\{U,V\}f)(D_1, \ldots, D_n) = Vf(UD_1, \ldots, UD_n).$

It becomes a 2-functor if, for natural transformations $\alpha\colon U \Rightarrow \bar{U}$ and $\beta\colon V \Rightarrow \bar{V}$ over $\underline{\underline{P}}$, we define $\{\alpha,\beta\}\colon \{U,V\} \Rightarrow \{\bar{U},\bar{V}\}$ as the natural transformation whose T-component is the horizontal composite

(2.32)

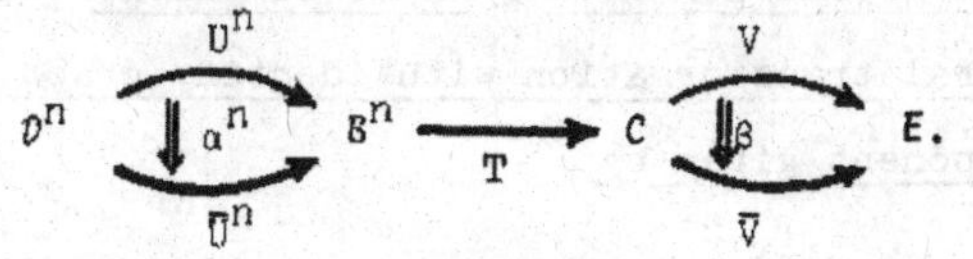

Note that (2.7) is the correct notation for this 2-functor if we take "op" to mean "reverse 1-cells but not 2-cells".

One word of warning: we have agreed to abbreviate the objects and morphisms of $\{B,C\}$ from (n,T) to T and from (ξ,f) to f; but the official definition must be remembered when, in degenerate cases, it might happen that natural transformations of different graphs ξ and η have the same components. For example, $I^\xi \colon I^n \to I^n$ is the identity functor for all ξ, so that when $B = I$ both (ξ,f) and (η,f) could make sense in the situation (2.22); they count as different morphisms of $\{I,C\}$. In particular $\{I,I\} = \underline{\underline{P}}$, and also $\{0,0\} = \underline{\underline{P}}$ where 0 is the empty category.

We now state the main result, asserting that $\underline{\underline{Cat}}/\underline{\underline{P}}$ is a closed 2-category; the proof is a straightforward verification which we leave to the reader:

<u>Theorem 2</u> <u>There is a 2-natural isomorphism of categories</u>

(2.33) $$\Pi\colon \underline{\underline{Cat}}/\underline{\underline{P}}\,(A\circ B,C) \cong \underline{\underline{Cat}}/\underline{\underline{P}}(A,\{B,C\})$$

<u>given as follows. For a functor</u> $U\colon A\circ B \to C$ <u>over</u> $\underline{\underline{P}}$, ΠU <u>is the functor</u> $A \to \{B,C\}$ <u>over</u> $\underline{\underline{P}}$ <u>given on objects by</u>

(2.34) $$((\Pi U)A)(B_1, \ldots, B_n) = U(A[B_1, \ldots, B_n])$$

(2.35) $$((\Pi U)A)(g_1, \ldots, g_n) = U(A[g_1, \ldots, g_n])$$

<u>and on morphisms by</u>

(2.36) $$((\Pi U)f)(B_1, \ldots, B_n) = U(f[B_1, \ldots, B_n]).$$

<u>For a natural transformation</u> $\alpha\colon U \Rightarrow \bar{U}$ over $\underline{\underline{P}}$, $\Pi\alpha\colon \Pi U \Rightarrow \Pi\bar{U}$ <u>is the natural transformation over</u> $\underline{\underline{P}}$ <u>whose component</u> $(\Pi\alpha)(A)\colon (\Pi U)A \to (\Pi\bar{U})A$ <u>is the natural transformation with identity graph and with</u> $(B_1, \ldots, B_n)$<u>-component given by</u>

(2.37) $\alpha(A[B_1, \ldots, B_n]) \colon U(A[B_1, \ldots, B_n]) \to \bar{U}(A[B_1, \ldots, B_n])$.

2.5 From this main result we now extract the generalization of the Godement calculus. The objects and the morphisms of $\{\mathcal{B},\mathcal{C}\}$ replace the functors and the natural transformations of the Godement calculus, and composition in $\{\mathcal{B},\mathcal{C}\}$ replaces "vertical" composition of natural transformations. By iterating the evaluation

(2.38) $$\varepsilon \colon \{\mathcal{B},\mathcal{C}\}\circ\mathcal{B} \to \mathcal{C}$$

corresponding to the adjunction (2.33), we get

$$\{\mathcal{B},\mathcal{C}\}\circ \{A,\mathcal{B}\}\circ A \xrightarrow{1\circ\varepsilon} \{\mathcal{B},\mathcal{C}\}\circ\mathcal{B} \xrightarrow{\varepsilon} \mathcal{C},$$

and hence by adjunction a functor over $\underline{\underline{P}}$

(2.39) $$\mu \colon \{\mathcal{B},\mathcal{C}\} \circ \{A,\mathcal{B}\} \to \{A,\mathcal{C}\}.$$

From the isomorphism $J\circ A \cong A$ we also get the functor over $\underline{\underline{P}}$

(2.40) $$\eta \colon J \to \{A,A\}.$$

These functors μ and η satisfy the associative and identity laws by the general theory of closed categories [2] (where they are called M and j).

η of course just sends the unique object 1 of J to the functor 1_A. If we write the images under μ of objects and of morphisms as

(2.41) $$\mu(T[S_1, \ldots, S_n]) = T(S_1, \ldots, S_n),$$
(2.42) $$\mu(f[g_1, \ldots, g_n]) = f(g_1, \ldots, g_n),$$

then (2.41) is the operation of substitution for functors, generalizing composition of functors in the ordinary Godement calculus, and (2.42) is the corresponding generalization of "horizontal" composition of natural transformations. In detail, the right side of (2.41) is the functor

$$(2.43)\quad A^m = A^{m_1} \times \dots \times A^{m_n} \xrightarrow[S_1 \times \dots \times S_n]{} B \times \dots \times B \xrightarrow[T]{} C$$

where $\Gamma T = n$ and $\Gamma S_i = m_i$, with $m = m_1 + \dots + m_n$; and the right side of (2.42) is the classical horizontal composite

$$(2.44)\quad A^{m_1} \times \dots \times A^{m_n} \;\underset{S'_1 \times \dots \times S'_n}{\overset{S_1A^{\eta_1} \times \dots \times S_nA^{\eta_n}}{\Downarrow g_1 \times \dots \times g_n}}\; B \times \dots \times B \;\underset{T'}{\overset{TB^{\xi}}{\Downarrow f}}\; C$$

As in the classical case, the general horizontal composite (2.42) can be expressed in terms of the special cases when either f or else the g_i are identities; for applying μ to (2.4) (after replacing A,B by T,S) gives the commutative diagram

$$(2.45)\quad \begin{array}{ccc} T(S_{\xi 1}, \dots, S_{\xi n}) & \xrightarrow{f(S_1, \dots, S_n)} & T'(S_1, \dots, S_n) \\ {\scriptstyle T(g_{\xi 1}, \dots, g_{\xi n})}\downarrow & \searrow^{f(g_1, \dots, g_n)} & \downarrow{\scriptstyle T'(g_1, \dots, g_n)} \\ T(S'_{\xi 1}, \dots, S'_{\xi n}) & \xrightarrow[f(S'_1, \dots, S'_n)]{} & T'(S'_1, \dots, S'_n). \end{array}$$

In terms of elements, $T(S_1, \dots, S_n)$ is the functor $A^m \to C$ given by

$$(2.46)\quad T(S_1,\dots, S_n)(A_1,\dots, A_m) = T(S_1(A_1, \dots, A_{m_1}),\dots, S_n(\dots,A_m))$$

and by a similar formula for morphisms; $T(g_1, \dots, g_n)$ is the natural transformation of graph $n(\eta_1, \dots, \eta_n)$ with components

$$(2.47)\quad T(g_1,\dots, g_n)(A_1,\dots, A_m) = T(g_1(A_1,\dots, A_{m_1}),\dots, g_n(\dots,A_m));$$

and $f(S_1, \dots, S_n)$ is the natural transformation of graph $\xi(m_1,\dots,m_n)$ with components

$$(2.48)\quad f(S_1,\dots, S_n)(A_1,\dots,A_m) = f(S_1(A_1,\dots, A_{m_1}),\dots, S_n(\dots,A_m)).$$

We end this section with the observation that $\mu\colon \{A,A\}\circ\{A,A\} \to \{A,A\}$ and $\eta\colon J \to \{A,A\}$ make of the "endomorphism object" $\{A,A\}$ a $\circ$- monoid. We shall argue in the following paper [5] that to give an extra structure on A, of the kind contemplated in §1.1, but with the natural transformations restricted to those of this paper, is precisely to give a $\circ$- monoid K in $\underline{\underline{Cat}}/\underline{\underline{P}}$ and a monoid-map $K \to \{A,A\}$; that in fact $\circ$- monoids are what coherence is all about.

The terminal object in any monoidal category has a unique monoid structure; hence $\underline{\underline{P}}$ is a $\circ$- monoid in $\underline{\underline{Cat}}/\underline{\underline{P}}$ (in fact $\underline{\underline{P}}$ can be identified with the endomorphism monoid $\{I,I\}$). Its multiplication $\mu\colon \underline{\underline{P}}\circ\underline{\underline{P}} \to \underline{\underline{P}}$ is the functor sending $n[m_1, \ldots, m_n]$ to $n(m_1, \ldots, m_n)$ and $\xi[\eta_1, \ldots, \eta_n]$ to $\xi(\eta_1, \ldots, \eta_n)$; its unit $\eta\colon J \to \underline{\underline{P}}$ sends $1 \in J$ to $1 \in \underline{\underline{P}}$.

3. The many-category case

3.1 We now indicate the way in which the calculus of §2 must be generalized to produce a calculus apt for the discussion of extra structures carried, no longer by one category, but by a family of categories.

An example of such a structure is a monoidal functor. Here we have categories A_1 and A_2; functors $\otimes_1\colon A_1{}^2 \to A_1$ and $I_1\colon A_1{}^0 \to A_1$, with natural transformations a_1 etc., making A_1 a monoidal category; functors $\otimes_2\colon A_2{}^2 \to A_2$ and $I_2\colon A_2{}^0 \to A_2$ with natural transformations a_2 etc., making A_2 a monoidal category; a functor $\phi\colon A_1 \to A_2$, and natural transformations $\tilde{\phi} : \phi A \otimes_2 \phi B \to \phi(A \otimes_1 B)$ and $\phi^o\colon I_2 \to \phi I_1$ satisfying appropriate axioms. The corresponding coherence problem has been discussed by Lewis ([10], in this volume); the easier case in which the identities I_1, I_2 were lacking was discussed earlier by Epstein [3].

A second example is that of two categories A_1, A_2, with a monoidal structure on A_1 given by functors $\otimes: A_1{}^2 \to A_1$ and $I: A_1{}^0 \to A_1$ with appropriate natural isomorphisms, and also a functor $\overline{\otimes}: A_1 \times A_2 \to A_2$ with natural isomorphisms $(A \otimes B)\overline{\otimes} C \to A \overline{\otimes}(B \overline{\otimes} C)$ and $I \overline{\otimes} C \to C$ subject to suitable axioms. The corresponding coherence problem was in effect solved by Epstein in §2 of [3] (without the identity I) and by Mac Donald [11] (with I); these authors really considered a different problem, with only one category but with mixed-variance functors, arising from the above by setting $A_2 = A_1{}^{op}$ - but their proofs apply unchanged to the present problem.

An example of a three-category problem, although one outside our present context in that the functors are _essentially_ of mixed variance and the natural transformations are those of [1], is provided by the Kelly - Mac Lane paper in this volume [8].

The carrier of such a structure is of course not a _set_ of categories - A_1 and A_2 could perfectly well coincide in our first example - but a _family_ of categories A_λ, $\lambda \in \Lambda$, for some indexing set Λ (which is usually finite, but could be arbitrary). Such a family (A_λ) may be denoted by a single letter A, and we shall call it a _polycategory_. A happier way to regard such a polycategory is to think of it as a category A _over_ Λ, treating the set Λ as a discrete category; if $\Delta: A \to \Lambda$ is the augmentation then A_λ is $\Delta^{-1}(\lambda)$. In the following discussion Λ is fixed, there being a separate calculus for each Λ.

In the one-category case a functor involved in the extra structure on A was a functor $A^n \to A$, and its _type_ was given by $n \in \underline{N}$. In the many-category case it will be of the form $A_{\lambda_1} \times A_{\lambda_2} \times \ldots{}_\wedge A_{\lambda_n} \to A_\mu$, and its _type_ must be specified by $(\lambda_1, \lambda_2, \ldots, \lambda_n; \mu)$. As in the one-category case, it turns out that the right things to consider are functors $T: A_{\lambda_1} \times \ldots \times A_{\lambda_n} \to B_\mu$

where A, B are polycategories (for the same Λ); these functors then form the objects of a generalized functor category $\{A,B\}$. The morphisms are again natural transformations f whose graph ξ is a permutation of n; but there are no such morphisms $T \to T'$ unless $\mu = \mu'$, $n = n'$, and $\lambda'_{\xi i} = \lambda_i$ for $i \in n$. When these conditions <u>are</u> satisfied, f has as before components $f(A_1, \ldots, A_n)$: $T(A_{\xi 1}, \ldots, A_{\xi n}) \to T'(A_1, \ldots, A_n)$ in B_μ, subject to the usual naturality condition in each A_i; this makes sense because $A_i \in A_{\lambda'_i}$ and hence $A_{\xi i} \in A_{\lambda_i}$. Then $\{A,B\}$ is a category over the category $\underline{\underline{Q}}$ of <u>types</u> and <u>graphs</u>; $\underline{\underline{Q}}$, which depends on Λ, has objects $(\lambda_1, \ldots, \lambda_n; \mu)$ and morphisms those permutations ξ satisfying the above conditions $\mu = \mu'$, $n = n'$, $\lambda'_{\xi i} = \lambda_i$.

Everything in §2 now carries over, with the category $\underline{\underline{Cat}}/\Lambda$ of polycategories replacing the category $\underline{\underline{Cat}}$ of categories, and with $\underline{\underline{Cat}}/\underline{\underline{Q}}$ replacing $\underline{\underline{Cat}}/\underline{\underline{P}}$; note that $\underline{\underline{Cat}}/\Lambda$ is a full subcategory of $\underline{\underline{Cat}}/\underline{\underline{Q}}$, just as $\underline{\underline{Cat}}$ is a full subcategory of $\underline{\underline{Cat}}/\underline{\underline{P}}$. We proceed to the details.

<u>3.2</u> The set Λ is fixed. The corresponding category $\underline{\underline{Q}}$ of types and graphs can be identified with the category $\underline{\underline{P}} \circ \Lambda \times \Lambda$, using $\circ$ in the sense of §2. For the object $(\lambda_1, \ldots, \lambda_n; \mu)$ of $\underline{\underline{Q}}$ can be identified with the object $(n[\lambda_1, \ldots, \lambda_n], \mu)$ of $\underline{\underline{P}} \circ \Lambda \times \Lambda$, and in future we write it in this way. Since Λ is discrete there are morphisms $(n[\lambda_1, \ldots, \lambda_n], \mu) \to (n'[\lambda'_1, \ldots, \lambda'_{n'}], \mu')$ in $\underline{\underline{P}} \circ \Lambda \times \Lambda$ only when $n = n'$ and $\mu = \mu'$; and then a morphism is necessarily of the form $(\xi[\lambda'_1, \ldots, \lambda'_n], \mu)$, where ξ is a permutation of n. For this to have the desired domain we must have

$$\lambda'_{\xi i} = \lambda_i \text{ for } i \in n; \tag{3.1}$$

so the morphisms of $\underline{\underline{P}} \circ \Lambda \times \Lambda$ are just those of $\underline{\underline{Q}}$. We shall normally abbreviate the morphism $\xi[\lambda'_1, \ldots, \lambda'_n]$ of $\underline{\underline{P}}\circ\Lambda$, or the morphism $(\xi[\lambda'_1, \ldots, \lambda'_n], \mu)$ of $\underline{\underline{P}}\circ\Lambda \times \Lambda$, to ξ, recalling that it satisfies (3.1).

The augmentation $\Gamma: A \to \underline{\underline{Q}}$ of a category over $\underline{\underline{Q}}$ is therefore given by functors $\Gamma': A \to \underline{\underline{P}}\circ\Lambda$ and $\Gamma'': A \to \Lambda$. We call $\Gamma'T$ the <u>domain type</u> of T and $\Gamma''T$ the <u>codomain type</u> of T. When Λ has just one element, $\underline{\underline{P}}\circ\Lambda$ becomes the category $\underline{\underline{P}}$, Λ becomes the unit category I, $\underline{\underline{Q}}$ becomes $\underline{\underline{P}}$, and we re-find the situation of §2 with $\Gamma = \Gamma'$ since Γ'' is trivial.

In §2 a category over $\underline{\underline{P}}$ is first of all a category; here a category $\Gamma: A \to \underline{\underline{Q}}$ over $\underline{\underline{Q}}$ is in particular a category over Λ, namely $\Gamma'': A \to \Lambda$. On the other hand, $\underline{\underline{Cat}}$ was regarded in §2 as a full subcategory of $\underline{\underline{Cat}}/\underline{\underline{P}}$; here $\underline{\underline{Cat}}/\Lambda$ becomes a full subcategory of $\underline{\underline{Cat}}/\underline{\underline{Q}}$ when we identify $\Delta: A \to \Lambda$ with $\Gamma: A \to \underline{\underline{Q}}$ where $\Gamma'' = \Delta$ and Γ' is the constant functor at $0[\ ]$.

The multiplication $\underline{\underline{P}}\circ\underline{\underline{P}} \to \underline{\underline{P}}$ described in §2.2 induces a functor $\underline{\underline{P}}\circ(\underline{\underline{P}}\circ\Lambda) = (\underline{\underline{P}}\circ\underline{\underline{P}})\circ\Lambda \to \underline{\underline{P}}\circ\Lambda$. We write this functor, on objects, as $n[\gamma_1, \ldots, \gamma_n] \mapsto n(\gamma_1, \ldots, \gamma_n)$ where $\gamma_i \in \underline{\underline{P}}\circ\Lambda$; so if $\gamma_1 = m_1[\lambda_1, \ldots, \lambda_{m_1}]$ etc., then

$$(3.2) \qquad n(\gamma_1, \ldots, \gamma_n) = n(m_1, \ldots, m_n)[\lambda_1, \ldots, \lambda_m]$$

where $m = m_1 + \ldots + m_n$. On morphisms, we also write it as $\xi[\eta_1, \ldots, \eta_n] \mapsto \xi(\eta_1, \ldots, \eta_n)$; happily this agrees with the meaning (2.15) of $\xi(\eta_1, \ldots, \eta_n)$ when a morphism of $\underline{\underline{P}}\circ\Lambda$ is identified with a permutation satisfying (3.1).

<u>3.3</u> For categories A, B over $\underline{\underline{Q}}$ we define a category $A\circ B$ over $\underline{\underline{Q}}$, generalizing the definition in §2.3; there is no danger of confusion in using the same symbol $\circ$, since Λ is supposed to be known.

An object of $A \circ B$ is $A[B_1, \ldots, B_n]$ where $A \in A$, $B_i \in B$, and where

(3.3) $\Gamma'A = n[\Gamma''B_1, \ldots, \Gamma''B_n]$;

thus the codomain types of the B_i are to match the domain type of A. A morphism is

(3.4) $f[g_1, \ldots, g_n] : A[B_1, \ldots, B_n] \to A'[B'_1, \ldots, B'_n]$

where $f: A \to A'$ with $\Gamma f = \xi$, and where $g_i : B_{\xi^{-1}i} \to B'_i$. Of course there are no such morphisms unless $\Gamma''A = \Gamma''A'$, for then there are no $f: A \to A'$; and moreover ξ is restricted as in (3.1). With composition defined as in (2.17), $A \circ B$ is a category; to augment it over $\underline{\underline{Q}}$ we set

(3.5) $\Gamma'(A[B_1, \ldots, B_n]) = n(\Gamma'B_1, \ldots, \Gamma'B_n)$,

(3.6) $\Gamma'(f[g_1, \ldots, g_n]) = \Gamma f(\Gamma g_1, \ldots, \Gamma g_n)$,

(3.7) $\Gamma''(A[B_1, \ldots, B_n]) = \Gamma''A$.

$T \circ S$ and $\alpha \circ \beta$ are defined as in (2.20) and (2.21), making $\circ$ a 2-functor $\circ$: $\underline{\underline{Cat}}/\underline{\underline{Q}} \times \underline{\underline{Cat}}/\underline{\underline{Q}} \to \underline{\underline{Cat}}/\underline{\underline{Q}}$. If $B \in \underline{\underline{Cat}}/\Lambda$, $A \circ B \in \underline{\underline{Cat}}/\Lambda$; for $n(0[\ ], 0[\ ], \ldots, 0[\ ]) = n(0, 0, \ldots, 0)[\ ] = 0[\ ]$.

As in §2.3, $\circ$ is coherently associative with a two-sided identity J; in the present case J is the category Λ, with augmentation Γ over $\underline{\underline{Q}}$ where $\Gamma'\lambda = 1[\lambda]$ and $\Gamma''\lambda = \lambda$.

<u>3.4</u> For categories B, C over $\underline{\underline{Q}}$ we define the category $\{B,C\}$ over $\underline{\underline{Q}}$. Set $B_\lambda = \Gamma''^{-1}\lambda$, and similarly for C_λ. An object of $\{B,C\}$ is a type $(n[\lambda_1, \ldots, \lambda_n], \mu)$ together with a functor $T: B_{\lambda_1} \times \ldots \times B_{\lambda_n} \to C_\mu$ satisfying the following analogues of (2.27) and (2.28):

(3.8) $\Gamma'(T(B_1, \ldots, B_n)) = n(\Gamma'B_1, \ldots, \Gamma'B_n)$

(3.9) $\Gamma'(T(g_1, \ldots, g_n)) = n(\Gamma g_1, \ldots, \Gamma g_n)$.

A morphism $T \to T'$ in $\{B,C\}$ exists only if there is a morphism $\xi: \Gamma T \to \Gamma T'$; that is, only if $n = n'$, $\mu = \mu'$, and (3.1) is satisfied. In this case a morphism $T \to T'$ is such an ξ together with a natural transformation $f: T \to T'$ of graph ξ whose components $f(B_1, \ldots, B_n)$: $T(B_{\xi 1}, \ldots, B_{\xi n}) \to T'(B_1, \ldots, B_n)$ satisfy the following analogue of (2.29):

(3.10) $\Gamma'(f(B_1, \ldots, B_n)) = \xi(m_1, \ldots, m_n)$

where $\Gamma' B_i = m_i[\nu_1, \ldots, \nu_{k_i}]$ for some ν_j. Note that (3.8) - (3.10) are automatically satisfied when $B, C \in \underline{\underline{Cat}}/\Lambda$. Finally $\{B,C\}$ is augmented over $\underline{\underline{Q}}$ by setting $\Gamma T = (n[\lambda_1, \ldots, \lambda_n], \mu)$ and $\Gamma f = \xi$.

$\{\ ,\ \}$ is made into a 2-functor by the appropriate generalizations of (2.31) and (2.32), which we leave the reader to formulate explicitly.

We then have:

<u>Theorem 3</u> <u>Theorem</u> 2 <u>continues to hold when $\underline{\underline{P}}$ is replaced throughout by $\underline{\underline{Q}}$</u>.

Everything in §2.5 carries over to the present situation, modulo some trivial notational changes. The functor $\eta: J \to \{A,A\}$ now sends $\lambda \in J$ to 1_{A_λ}; (2.43) and (2.44) need re-writing, but otherwise (2.41) - (2.48) stand. Now $\{A,A\}$ is a $\circ$- monoid in $\underline{\underline{Cat}}/\underline{\underline{Q}}$; the terminal object $\underline{\underline{Q}}$ of this category is itself a $\circ$- monoid, identifiable with $\{\Lambda,\Lambda\}$. This map $\underline{\underline{Q}}\circ\underline{\underline{Q}} \to \underline{\underline{Q}}$ sends $(n[\lambda_1, \ldots, \lambda_n], \mu)\,[(m_1[\nu_1, \ldots], \lambda_1), \ldots, (m_n[\ldots, \nu_k], \lambda_n)]$ to $(n(m_1, \ldots, m_n)\,[\nu_1, \ldots, \nu_k], \mu)$ and $\xi[\eta_1, \ldots, \eta_n]$ to $\xi(\eta_1, \ldots, \eta_n)$; the map $J \to \underline{\underline{Q}}$ sends λ to $(1[\lambda], \lambda)$.

<u>3.5</u> Finally we consider the relation between the calculi for

various Λ; to indicate the dependence on Λ we write $\mathbf{Q}_\Lambda$, $\circ_\Lambda$, $\{\ ,\ \}_\Lambda$, J_Λ for $\mathbf{Q}$, $\circ$, $\{\ ,\ \}$, J.

Any map $k\colon \Lambda \to M$ of index sets induces a functor $\mathbf{P}\circ\Lambda \times \Lambda \to \mathbf{P}\circ M \times M$, that is, $\mathbf{Q}_\Lambda \to \mathbf{Q}_M$; and hence a 2-functor $\psi\colon \mathbf{Cat}/\mathbf{Q}_\Lambda \to \mathbf{Cat}/\mathbf{Q}_M$. This has a right adjoint ϕ sending $\Delta\colon A \to \mathbf{Q}_M$ to $\Gamma\colon \phi A \to \mathbf{Q}_\Lambda$, where ϕA and Γ are defined by the pullback

$$\begin{array}{ccc} \phi A & \longrightarrow & A \\ {\scriptstyle \Gamma}\downarrow & & \downarrow{\scriptstyle \Delta} \\ \mathbf{Q}_\Lambda & \longrightarrow & \mathbf{Q}_M \end{array} \tag{3.11}$$

The functor ψ is anti-monoidal; that is, it is a monoidal functor $(\mathbf{Cat}/\mathbf{Q}_\Lambda)^{op} \to (\mathbf{Cat}/\mathbf{Q}_M)^{op}$; the maps $\psi(A\circ_\Lambda B) \to \psi A \circ_M \psi B$ and $\psi J_\Lambda \to J_M$ providing its extra structure are evident. When $k\colon \Lambda \to M$ is the inclusion of a non-empty subset, that is when it is a coretraction, it is clear that ψ embeds $\mathbf{Cat}/\mathbf{Q}_\Lambda$ as a full subcategory of $\mathbf{Cat}/\mathbf{Q}_M$, and preserves $\circ$ (but not J).

The functor ϕ is monoidal, with evident maps $\phi A \circ_\Lambda \phi B \to \phi(A\circ_M B)$ and $\phi J_M \to J_\Lambda$. When k is an inclusion ϕ preserves J (but not $\circ$). Since ϕ is monoidal, it takes a $\circ_M$- monoid to a $\circ_\Lambda$- monoid, and an action of the first on A to an action of the second on ϕA.

4. More general natural transformations

4.1 It suffices to consider the one-category case, the extension to many categories presenting no problems. We consider first more general natural transformations in which the functors are still covariant; so a functor is of the form $T\colon A^n \to B$ and its type is given by $n \in \mathbf{N}$.

Write $\mathbf{S}$ for the skeletal category of finite sets with $\mathbf{N}$ as its set of objects and with functions $n \to m$ as morphisms; our category $\underline{\underline{P}}$ is the subcategory of $\underline{\underline{S}}$ in which only the isomorphisms of $\underline{\underline{S}}$ are retained.

For $\phi: n \to m$ in $\underline{\underline{S}}$ and for $T: A^n \to B$, $S: A^m \to B$, we define a natural transformation $f: T \to S$ of graph ϕ to be a classical natural transformation $f: TA^{\phi} \Rightarrow S$, generalizing (2.22); it has components $f(A_1, \ldots, A_m): T(A_{\phi 1}, \ldots, A_{\phi n}) \to S(A_1, \ldots, A_m)$. Everything above extends fairly immediately to this case: we get a new generalized functor category $\{B,C\}$ over $\underline{\underline{S}}$, with left adjoint $A \circ B$ in which a morphism $A[B_1, \ldots, B_n] \to A'[B_1', \ldots, B_m']$ is $f[g_1, \ldots, g_n]$ where $\Gamma f = \phi$ and where $g_i: B_i \to B'_{\phi i}$.

Such natural transformations occur, for instance, when we are describing the extra structure on A consisting in admitting finite coproducts. The basic functors are $+ : A^2 \to A$ and $I: A^0 \to A$; the basic natural transformations are $p: A \to A+B$, $q: B \to A+B$, $d: A+A \to A$, $i: I \to A$, of respective graphs $\ulcorner 1 \urcorner: 1 \to 2$, $\ulcorner 2 \urcorner: 1 \to 2$, the unique function $2 \to 1$, and the unique function $0 \to 1$. To ensure that $+$ really is the coproduct and I the initial object, we have to impose the purely equational axioms saying that each of the following is the identity:

$$A \xrightarrow{p} A+A \xrightarrow{d} A \qquad A \xrightarrow{q} A+A \xrightarrow{d} A$$

$$A+B \xrightarrow{p+q} (A+B)+(A+B) \xrightarrow{d} A+B \qquad I \xrightarrow{i_I} I.$$

The ideas of the following paper [5] also extend at once to this case; in particular the free category-with-finite-coproducts on a given category B is $\underline{\underline{S}} \circ B$.

If we want the kind of natural transformation needed in discussing finite products, we must take those with graph in $\underline{\underline{S}}^{op}$; if

$T: A^n \to B$ and $S: A^m \to B$, a natural transformation $f: T \to S$ of graph $\psi: m \to n$ is a classical natural transformation $f: T \Rightarrow SA^{\psi}$. Once again there is a functor category $\{B,C\}$ over $\underline{\underline{S}}^{op}$, with left adjoint $A \circ B$, a morphism in the latter now being $f[g_1, \ldots, g_m]: A[B_1, \ldots, B_n] \to A'[B'_1, \ldots, B'_m]$ where $\Gamma f = \psi$ and $g_i: B_{\psi i} \to B'_i$. Again the ideas of the following paper extend to this case.

If we want to talk about products and coproducts together, we should need more general natural transformations, with components like the composites $A+A \to A \to A \times A$ and $A \times B \to A \to A+C$. Similarly if we want to talk about a distributive law in a category: we have natural transformations $d: A \otimes (B \oplus C) \to (A \otimes B) \oplus (A \otimes C)$ and $d^{-1}: (A \otimes B) \oplus (A \otimes C) \to A \otimes (B \oplus C)$. The graph, telling which arguments are to be set equal, now has to consist of two functions

$$n \xrightarrow{\phi} k \xleftarrow{\psi} m, \qquad (4.1)$$

and the components are of the form $f(A_1, \ldots, A_k): T(A_{\phi 1}, \ldots, A_{\phi n}) \to S(A_{\psi 1}, \ldots, A_{\psi m})$. The category $\underline{\underline{G}}$ of types and graphs has $\underline{\underline{N}}$ as its set of objects, and a morphism $\xi: n \to m$ is a diagram in $\underline{\underline{S}}$ like (4.1), different diagrams counting as the same graph if they differ only by an automorphism (= permutation) of k. We define the composite in $\underline{\underline{G}}$ of $\xi = (4.1)$ and $\eta: m \to p \leftarrow t$ to be the graph $\eta\xi: n \to q \leftarrow t$ got by forming the diagram

$$\begin{array}{ccccc} & & & & t \\ & & & & \downarrow \\ & & m & \longrightarrow & p \\ & & \downarrow & & \downarrow \\ n & \longrightarrow & k & \longrightarrow & q \end{array} \qquad (4.2)$$

in which the square is a pushout. For $T: A^n \to B$ and $S: A^m \to B$ a natural transformation $f: T \to S$ of graph ξ is now a classical natural transformation $f: TA^{\phi} \Rightarrow SA^{\psi}$, thus:

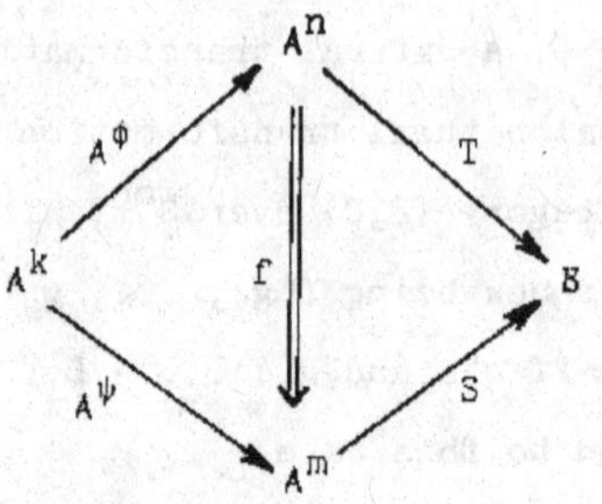

If g: S → R has graph η, the composite gf: T → R of graph ηξ is defined as the composite 2-cell

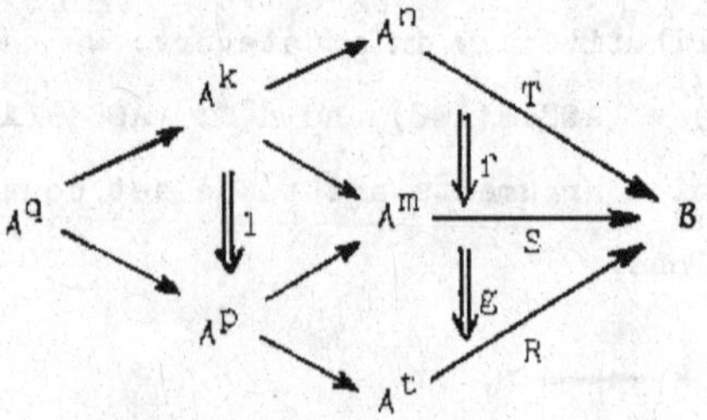

This certainly produces a functor category {B,C} over $\underline{\underline{G}}$. Moreover its evident continuity in C ensures that a left adjoint A∘B will exist. However the general morphism of A∘B is quite complicated, and it is not yet clear how far this provides a good context for discussing coherence problems. The complication is surely in the nature of things: the free category-with-finite-products-and-finite-coproducts on one generator is already pretty complicated.

Moreover there is a new aspect that arises here: if f: T → S is a natural transformation of graph (4.1), and if θ: k → h, there is a natural transformation of graph

$$n \xrightarrow{\theta\phi} h \xleftarrow{\theta\psi} m,$$

namely the classical composite fA^θ. This operation does occur in respectable problems; take the "distributive law" situation above, and write d as d: $T \to S$, with $d^{-1}: S \to T$; if the graphs of d and d^{-1} are ξ and η, then $\eta\xi = 1$ but $\xi\eta \neq 1$; and dd^{-1} is indeed not the identity natural transformation of S but instead that of some SA^{θ}, with components $1: (A\otimes B)\oplus(A\otimes C) \to (A\otimes B)\oplus(A\otimes C)$. This casts doubt on the wisdom of seeking a calculus with no explicit mention of functors A^{θ}; yet it is so obviously right in the simpler cases, where both coherence considerations and free structures are made much easier by not having TA^{θ} alongside T, that it is presumably worth looking for such a calculus in the general case.

Finally, observe that one further generalization is evidently called for: if we want to demand of a category that it admit equalizers or countable products or other limits, we can do so in the present context provided we replace n, m, k in (4.1) by small categories, and ϕ,ψ by functors. In the cases like $\underline{\underline{S}}$ or $\underline{\underline{S}}^{op}$, where either ϕ or ψ is 1, everything still works much as in the present paper.

4.2 We now come to the mixed-variance case. Here (in the one-category calculus) a functor is of the form $T: A\times A^{op}\times A^{op}\times A\times A \to B$, and its type is given by a string $\nu = (+ -- ++)$ of +'s and -'s indicating the variances. With ν as above we might write A^{ν} for $A\times A^{op}\times A^{op}\times A\times A$, so that $T: A^{\nu} \to B$; and write $n = |\nu|$ for the total number of arguments (5 in the above case). In fact a type can be considered as an object $5[+ -- ++]$ of $\underline{\underline{P}}\circ 2$, where 2 is the discrete category with objects $\{+,-\}$.

The natural transformations may be of various levels of complexity, corresponding for example to the categories $\underline{\underline{P}}$, $\underline{\underline{S}}$, $\underline{\underline{S}}^{op}$, $\underline{\underline{G}}$ above. The ones corresponding to $\underline{\underline{P}}$ are those considered in [1], and we take this easier case first. For types μ and ν, write $-\mu+\nu$ for the type of length $|\mu|+|\nu|$ got by writing first the string μ with all the signs changed and then the string ν. A graph $\xi: \mu \to \nu$ is then a bijection between +'s and the -'s in $-\mu + \nu$. If we write μ and ν as columns it can be written as an actual geometrical "graph" as in

(4.3) 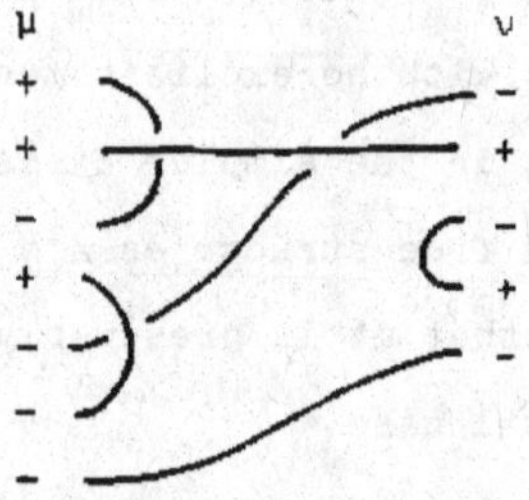

Composition of graphs is defined by joining them at the middle type where they meet; the composite of $\xi: \mu \to \nu$ as in (4.3) and $\eta: \nu \to \tau$ as in

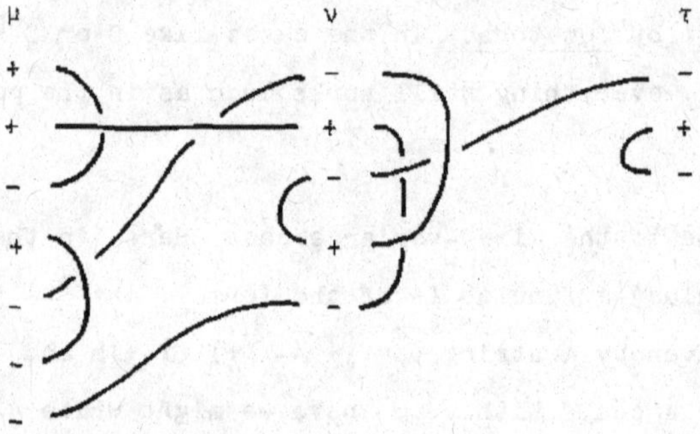

is the following graph $\eta\xi$

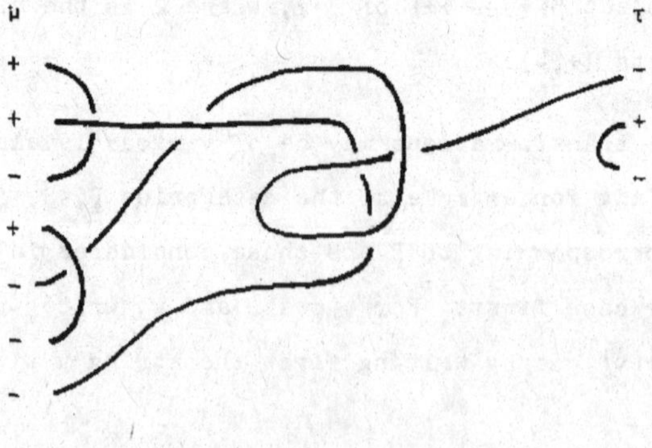

whose "linkages" may now be straightened out to taste. The problem, however, is that upon composing we may get closed loops, as in

(4.4)

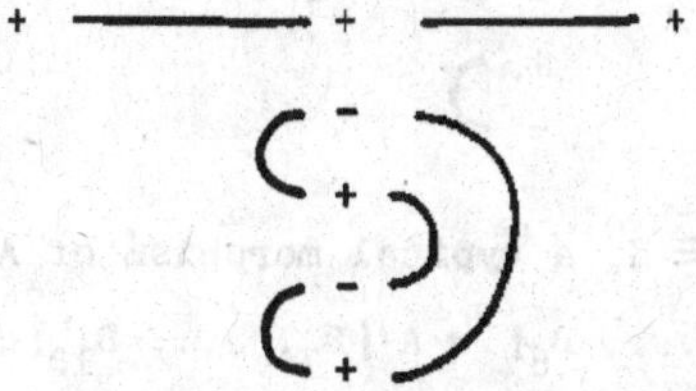

In such cases the graphs ξ and η were said in [1] to be <u>incompatible</u>; the composite was still defined there, namely as the part of (4.4) remaining after discarding the closed loops. We then get a category of types and graphs that we shall call $\underline{\underline{P}}_0^*$. A different possibility is to <u>allow</u> graphs containing closed loops; then we may get more closed loops upon composition, as in

(4.5)

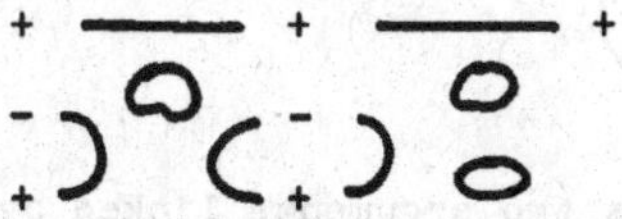

In this way we would seem to get a category, which we shall call $\underline{\underline{P}}^*$; the original graphs without closed loops can then be called <u>simple</u> graphs.

For a simple graph $\xi\colon \mu \to \nu$, the concept of a natural transformation $f\colon T \to S$ of graph ξ, where $T\colon A^{\mu} \to B$ and $S\colon A^{\nu} \to B$, was defined in [1]; it was moreover shown how to compose f with $g\colon S \to R$ of graph η to get a composite natural transformation gf of graph $\eta\xi$ when η and ξ were compatible.

As long as the graphs are simple, the calculus of this paper seems to extend; if A is a category over $\underline{\underline{P}}_0^*$ and B is a category, let $f\colon A \to A'$ have graph $\xi\colon \mu \to \nu$ given by

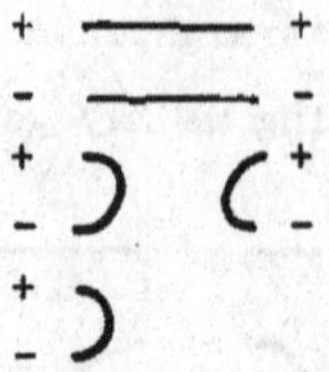

Then for $B_1, \ldots, B_{10} \in \mathcal{B}$, a typical morphism of $A \circ \mathcal{B}$ is $f[g_1, \ldots, g_5] : A[B_1, \ldots, B_6] \to A'[B_7, \ldots, B_{10}]$ where the maps g_i in $\mathcal{B}$ go like this:

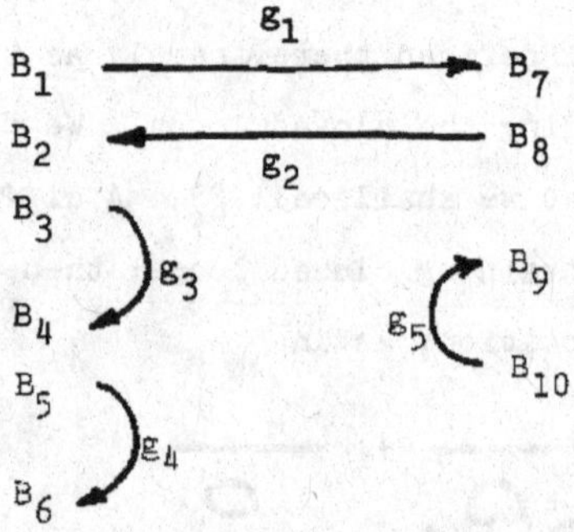

That is, $g: B_i \to B_j$ joins two arguments linked by ξ, and goes *from* the one which is *contravariant* in $-\mu +\nu$ *to* the one that is *covariant* in $-\mu +\nu$. It is moreover clear how to compose such morphisms in $A \circ \mathcal{B}$ as long as the graphs are *compatible*. Moreover substitution presents no problems: our earlier operation $\xi(\eta_1, \ldots, \eta_n)$ on graphs carries over well; if

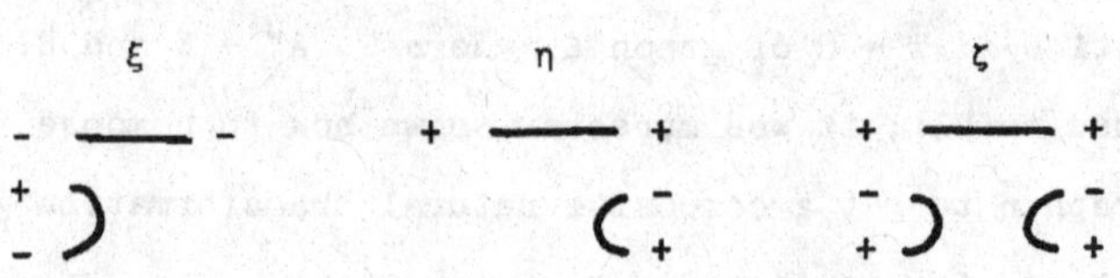

then $\xi(\eta\ , \zeta)$ is the result of replacing the linkages in ξ (under a microscope, as it were) by the graphs η and ζ:

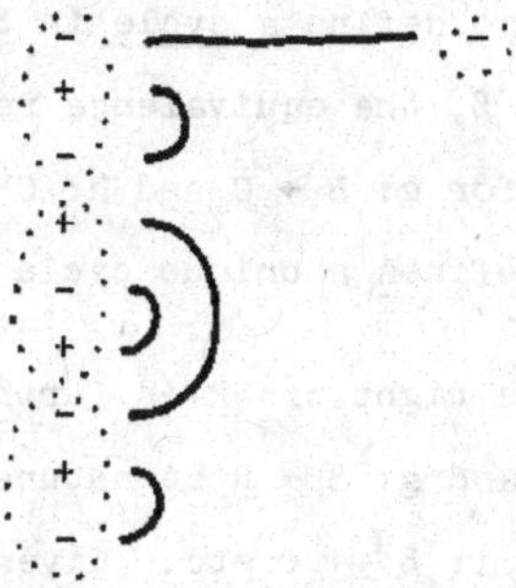

The problem is that there is just no {B,C} if we stick to simple graphs, for we cannot compose natural transformations of incompatible graphs; and correlatively that we cannot compose in A∘B if the graphs are incompatible. This suggests that we should allow the more general graphs of $\underline{P}^{\#}$. Then when we try to compose in A∘B, we get something fairly sensible. If f and k have graphs as in (4.4), a composite in A∘B looks something like

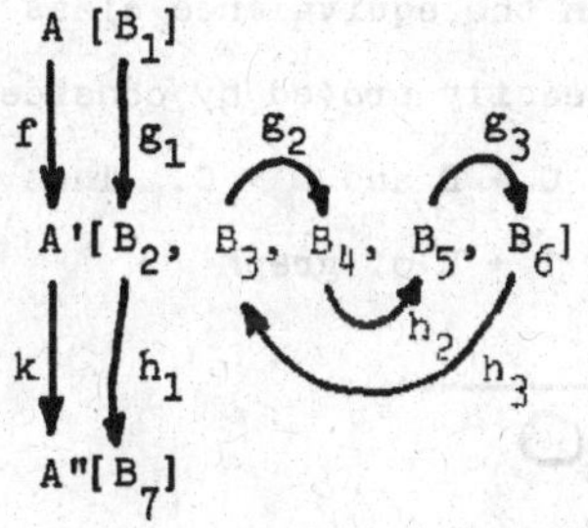

This suggests that a morphism in A∘B of the composite graph (4.4) should have a map $h_1 g_1 : B_1 \to B_7$ corresponding to the simple part of the graph, and something like

(4.6)

$$
\begin{array}{ccc}
B_3 & \xrightarrow{g_2} & B_4 \\
h_3 \uparrow & & \downarrow h_2 \\
B_6 & \xleftarrow{g_3} & B_5
\end{array}
$$

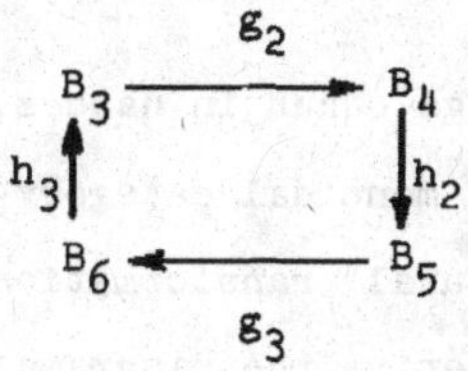

corresponding to the closed loop. There is no way of knowing where to start the loop (4.6), so define a <u>cycle</u> in $\mathcal{B}$ as an equivalence class of endomorphisms in $\mathcal{B}$, the equivalence relation being generated by the requirement that, for $g: B \to C$ and $h: C \to B$, gh and hg are equivalent. Then (4.6) defines a unique cycle in $\mathcal{B}$.

That we are on the right track is further suggested by the following. Let $f: T \to S$ and $g: S \to R$ be natural transformations with graphs as in (4.4), where $T': \mathcal{B}^{\mu} \to \mathcal{C}$ etc. Given B in $\mathcal{B}$ and $h: C \to C$ in $\mathcal{B}$, the composites

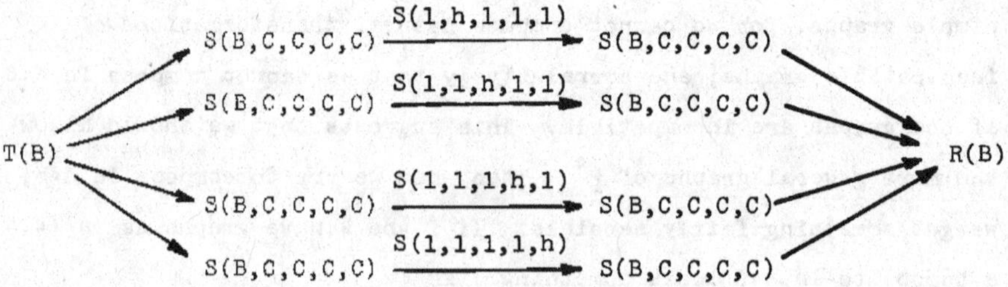

are all equal and depend only on the equivalence class [h] of h; the general result of this kind is easily proved by considering slightly more general diagrams with maps $C \to D$ and $D \to C$. Thus we are led to define a natural transformation $T \to R$ of graph

as consisting of components $f(B, \phi): T(B) \to R(B)$, natural in B and with ϕ a cycle of $\mathcal{B}$. This doesn't seem a bad idea, since it is the same thing as a natural transformation $\mathcal{B}(C,C) \to \mathcal{C}(TB,RB)$ in the sense of [1].

Moreover such things occur in nature. Define a <u>compact closed category</u> to be a symmetric monoidal category $(\mathcal{A}, \otimes, I)$ together with a functor $^*: \mathcal{A}^{op} \to \mathcal{A}$ and natural transformations $g_A: I \to A \otimes A^*$, $h_A: A^* \otimes A \to I$ making commutative the diagrams

$$\begin{array}{ccc} A & \xrightarrow{g\otimes 1} & A\otimes A^*\otimes A \\ & {\scriptstyle 1}\searrow & \downarrow{\scriptstyle 1\otimes h} \\ & & A \end{array} \qquad\qquad \begin{array}{ccc} A^* & \xrightarrow{1\otimes g} & A^*\otimes A\otimes A^* \\ & {\scriptstyle 1}\searrow & \downarrow{\scriptstyle h\otimes 1} \\ & & A^* \end{array}$$

(This is deliberately an equational definition; the mere existence of A^*, adjoint to A in the degenerate 2-category A, implies its functoriality and the naturality of g and h.) Such a category is closed, with $[A,B] = A^*\otimes B$; conversely a closed category is compact exactly when the canonical map $k: A\otimes[A,I] \to [A,A\otimes I]$ is an isomorphism, whereupon $A^* = [A,I]$, h is the evaluation, and g is $I \to [A,A] \cong [A,A\otimes I]$ followed by k^{-1}. Examples are the finite-dimensional vector spaces over a field I, or more generally the finite-dimensional representations over I of a group G. In such a category the composite

$$(4.7) \qquad I \xrightarrow{cg} A^*\otimes A \xrightarrow{h} I$$

is multiplication by dim A; we however would take the composite of the incompatibles cg and h to be

$$(4.8) \qquad I \xrightarrow{cg} A^*\otimes A \xrightarrow{u^*\otimes v} B^*\otimes B \xrightarrow{h} I,$$

which is multiplication by trace (uv) = trace (vu), depending on the cycle [uv] = [vu].

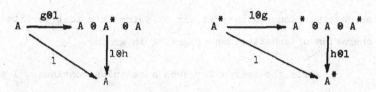

We seem, therefore, to be tantalizingly close to the right calculus, defined on categories over $\underline{\underline{P}}^*$. The trouble is that I don't really know how to define $\underline{\underline{P}}^*$. It's all very well to draw pictures like (4.5), but I don't see how to make real sense of the above unless we can order the closed loops in a graph so as to tell one from the other. It then transpires that, to get an order on the composite graph, we must in the factors have ordered the closed loops not only among themselves but also with respect to the non-closed linkages. I haven't yet found a natural way of doing this that permits any decent operation of substitution of graphs in graphs.

While the search for such a calculus continues, I can only use an emasculated form, defining $A \circ B$ for categories over $\underline{\underline{P}}_0^*$ such that composable maps in A or in B never have incompatible graphs. At least such categories occur often; it is proved in [7], [8], and [10] that incompatibilities do not occur in the problems there studied, and in a later paper [6] in this volume I show that this is always so when the only contravariant functors in the structure arise by positing adjoints for some of the covariant ones. The emasculated calculus, therefore, does serve as a stop-gap measure for describing things like free closed categories explicitly, and we use it in the following paper [5].

4.3 If a good $\underline{\underline{P}}^*$ is found, all will surely be well with $\underline{\underline{S}}^*$ and $\underline{\underline{S}}^{op*}$. At the moment I can define an $\underline{\underline{S}}_0^{op*}$: a graph is not a bijection of the +'s with the −'s in $-\mu + \nu$, but a function from the +'s to the −'s. Graphs can be composed if they are compatible in a suitable sense; I think they are always compatible in the theory of a cartesian closed category, and that the "coherence result" of Szabo ([13], [14]) for these is probably equivalent to the much simpler assertion that, if A is the free such on one generator, then $\Gamma: A \to \underline{\underline{S}}_0^{op*}$ is faithful. However I have yet to verify this, and I know still less about more complicated cases like $\underline{\underline{G}}^*$. If the following paper shows these ideas to be worthwhile, perhaps some colleague will supply my lack of wit.

REFERENCES

[1] S. Eilenberg and G.M. Kelly, A generalization of the functorial calculus, J. Algebra 3(1966), 366-375.

[2] S. Eilenberg and G.M. Kelly, Closed Categories, in: Proc. Conf. on Categorical Algebra, La Jolla, 1965 (Springer-Verlag, 1966), 421-562.

[3] D.B.A. Epstein, Functors between tensored categories, Invent. Math. 1(1966), 221-228.

[4] J.W. Gray, The categorical comprehension scheme, Lecture Notes in Mathematics 99(1969), 242-312.

[5] G.M. Kelly, An abstract approach to coherence. (in this volume).

[6] G.M. Kelly, A cut-elimination theorem. (in this volume).

[7] G.M. Kelly and S. Mac Lane, Coherence in closed categories, J. Pure and Applied Algebra 1(1971), 97-140.

[8] G.M. Kelly and S. Mac Lane, Closed coherence for a natural transformation. (in this volume).

[9] F.W. Lawvere, Functorial semantics of algebraic theories, Proc. Nat. Acad. Sci. U.S.A. 50(1963), 869-872.

[10] G. Lewis, Coherence for a closed functor. (in this volume).

[11] J.L. Mac Donald, Coherence of adjoints, associativities, and identities, Arch. Math. 19(1968), 398-401.

[12] S. Mac Lane, Natural associativity and commutativity, Rice University Studies 49(1963), 28-46.

[13] M.E. Szabo, Proof-theoretical investigations in categorical algebra (Ph. D. Thesis, McGill Univ., 1970).

[14] M.E. Szabo, A categorical equivalence of proofs (to appear).

AN ABSTRACT APPROACH TO COHERENCE. *

G.M. Kelly

The University of New South Wales, Kensington 2033, Australia

Received May 22, 1972

1. Introduction

1.1 We assume familiarity with the preceding paper [5] in this volume, and we refer the reader again to §1.1 of that paper.

Suppose we have a category (or polycategory) A provided with an extra structure of the kind considered there. So we are given basic functors, of the form $A^n \to A$ in the one-category fully-covariant case, and of the appropriate more general form in the many-category mixed-variance case. We pass at once to the wider class of allowable functors $T: A^n \to A$ obtained from these and the identity functor by iterated substitution. We are then given basic natural transformations $T \to S$ between certain pairs of these, each such having a graph as in [5] describing the arguments to be set equal. We now pass to the wider class of allowable natural transformations, obtained from the basic ones by adding the identities 1_T and allowing unlimited composition and substitution. If $f, g: T \to S$ are two such with the same domain and codomain, it makes no sense to ask whether $f = g$ unless they have the same graphs $\Gamma f = \Gamma g$; otherwise the diagram of their components is not closed. There is now given a set of axioms requiring certain formally-different pairs $f, g: T \to S$ of the same graph to coincide.

* Preliminary research for this paper was supported by grants from the National Science Foundation and the Louis Block Fund of the University of Chicago.

1.2 The coherence problem for the given structure is that of deciding which other formally-different pairs $f,\ g: T \longrightarrow S$ of allowable natural transformations necessarily coincide as a consequence of the axioms.

In the first examples, such as those considered by Mac Lane in [14], the problem was posed rather differently: there it was a question of *finding* a suitable finite set of axioms $u = v: P \to Q$ which would imply that *all diagrams commute*, that is, that $f,\ g: T \to S$ *always* coincide when $\Gamma f = \Gamma g$. The problem was so posed precisely because the natural examples of monoidal categories and similar structures *do* have all diagrams commuting.

In such structures as closed categories, however, it is just not the case that all diagrams commute in the natural examples. When studying coherence for closed categories in [7], Kelly - Mac Lane contented themselves therefore with proving a partial result of the form: $f,\ g: T \to S$ coincide if $\Gamma f = \Gamma g$ and if T, S belong to a certain *subset* of the allowable functors (there called the *proper* ones).

My student G. Lewis attacked the problem of extending the above result to the case of two closed categories connected by a closed functor $\phi,\ \tilde{\phi},\ \phi^{o}: A \to A'$. Here the first step was to deal with the case of a monoidal functor; this had been done by Epstein [3], with an "all diagrams commute" result, except that his A, A' lacked identities I, I' and so the corresponding $\phi^{o}: I' \to \phi I$ was missing. When the identities are allowed for, there are again non-commuting diagrams, *even in this purely covariant case*: the diagram

$$\begin{array}{ccc} \phi I & \xrightarrow{\ \cong\ } & I' \otimes \phi I \\ {\scriptstyle \cong}\downarrow & & \downarrow{\scriptstyle \phi^{o} \otimes 1} \\ \phi I \otimes I' & \xrightarrow[\ 1 \otimes \phi^{o}\ ]{} & \phi I \otimes \phi I \end{array}$$

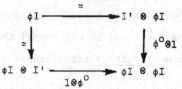

fails to commute for the forgetful monoidal functor **Ab** $\to$ **Set**. (Note that once again, as in [7], the constants turn out to be the villains which destroy commutativity.) Here again, a partial result like that of [7] could be given: f, g: T $\to$ S coincide if $\Gamma f = \Gamma g$ and if S is of the form ϕR. However Lewis went further, and gave necessary and sufficient conditions for f = g when T,S were arbitrary. This is what we could call a full solution of the coherence problem, as posed above in the first sentence of §1.2. The results of Lewis, re-formulated now in the setting developed in this paper, will be found in [12] in this volume.

1.3 This brings us to the question of a suitable formulation of the general coherence problem. Much that we have said above is fairly loose. We spoke of the allowable functors and the allowable natural transformations as if they formed a subcategory **All** = **All**(A) of the generalized functor category {A,A} of [5]; then an "all diagrams commute" result was the assertion that the restriction to **All** of the graph-functor Γ was faithful. But the definition of **All** is faulty: for two formally different allowable functors Q, T: $A^n \to A$ might, for this particular model A, fortuitously coincide. Then honest allowable natural transformations f: P $\to$ Q, g: T $\to$ S could be composed in {A,A} to form a freak "allowable" gf not writable for a general model A. This difficulty, overlooked in [14], was explicitly recognized in [7], where the objects of **All** were taken to be the formally different allowable functors (there called shapes), and not their realizations in {A,A}. (We have so far ignored for simplicity the fact that some theories require certain iterates of the basic functors to coincide, as a strict monoidal category has $(-\otimes-)\otimes- = -\otimes(-\otimes-)$; we shall allow for these in our formal theory below; they are to be distinguished from fortuitous coincidences.)

Even in [7], however, the morphisms in **All** were actual

natural transformations in $\{A,A\}$ between the functors corresponding to the shapes. This is satisfactory where, as there, we seek only a positive result of the form "$f,g: T \to S$ coincide if $\Gamma f = \Gamma g$ and T,S are proper"; it will no longer do when we seek <u>necessary and sufficient</u> conditions for $f = g$, for we may have $f = g$ in the All = All(A) for the for the model A, without the formally-corresponding f and g coinciding in All(B) for a general model B. We cannot even make sense of what we are trying to say here without abandoning talk of models and talking instead of the corresponding <u>theory</u>. This leads us to a <u>universal All</u> whose objects are the shapes and whose morphisms are <u>formal</u> allowable natural transformations; then $f = g$ in All if and only if their realizations coincide in All(A) for all models A.

The primary object of this paper is to formulate coherence problems in these terms, making clear the relation between the theory and its models. I wish to re-iterate my debt to Lewis, whose insistence that we should always try for a "full solution" set me looking for a suitable formulation.

<u>1.4</u> The universal All above (we shall now drop this notation) is a category K over the appropriate category $\underline{G}$ of types and graphs; moreover, since it contains a formal identity functor 1 and admits formal substitution, it is a $\circ$- monoid in the category $\underline{Cat}/\underline{G}$. The basic data and axioms provide a system of <u>generators and relations</u> for K, and in this view the coherence problem consists in calculating K from its generators and relations.

In the first problems studied, as in [14],[3], [13], commutativity of a diagram in K was established by showing that it could be "filled in" with a trellis of little diagrams known to commute - these little diagrams being instances of the axioms, or of "functoriality" conditions like $(f\otimes 1)(1\otimes g) = (1\otimes g)(f\otimes 1)$, or else of naturality conditions for the basic natural transformations. In later studies such as that of [7], where a cut-elimination technique was used, the methods used to establish commutativity made no explicit reference to "filling in". An important theoretical conclusion from our description of K in terms of generators and relations is that all commutative diagrams do in fact commute in virtue of just such a "filling in".

The $\circ$- monoid K in Cat/Q contains all the information necessary to describe the particular kind of extra structure in question, just as much as does the list of basic data and axioms; in fact to give such a structure to the category (or polycategory) A is precisely to give a map $\phi: K \to \{A,A\}$ of $\circ$- monoids. (Our earlier All(A) is roughly the image of ϕ; same objects as K, but morphisms equivalence classes of those in K, two being equivalent if they have the same image under ϕ.) To give such a ϕ is alternatively to give an <u>action</u> $\theta: K\circ A \to A$, exhibiting A as an algebra for the monad (= triple) $K\circ-$ on Cat (or on the category Cat/Λ of polycategories). This shows that the theory specified by K is a <u>doctrine</u> in the sense of Lawvere [11] (for $K\circ-$ is actually a 2-monad); of course it is a doctrine of a very special kind.

A knowledge of K carries with it an explicit knowledge of the <u>free structures</u>; the free structure on A is just $K\circ A$. Thus we are able to relate the coherence problem to the free-structure problem as studied by Lambek ([9],[10]) and Szabo ([15],[16], [17]). That some relation exists has been more or less vaguely realized for a long time; the history as I know it seems to be as follows. The Corollary to Lemma 5 in Lambek's paper [9] on free residuated categories, which says something like "any two proofs of T → S are equivalent if each variable appears at most twice", was highly reminiscent of the statement of the corresponding coherence problem in terms of the every-variable-twice generalized natural transformations of [1]. I remarked to Lambek that this result would most likely be seen, upon examination, to contain an "all diagrams commute" coherence result; and in his next paper, on free biclosed categories, Lambek used the name "coherence theorem" for the main result "two proofs with the same generality are equivalent"; Szabo, then Lambek's student, later used the same name in his thesis [15], where he discussed free closed categories and free cartesian closed categories. Meanwhile Mac Lane and I, independently of Szabo, drew upon Lambek's brilliant generalization of Gentzen's cut-elimination theorem to prove a partial coherence result for closed categories [7], observing that an "all diagrams commute" result was false; but even then we had not any clear notion of the relation of Lambek's main result to coherence as we understood it - nor I think did anyone else. The present work, expressing the free category-with-structure as $K \circ A$, finally makes the connexion clear; it turns out that Lambek's "main result" indeed says precisely that "all diagrams commute" - and is therefore in fact false for biclosed categories. The corresponding result for closed categories in Szabo's thesis is similarly false; his paper [17] contains a modified result asserted to give necessary and sufficient conditions for equivalence of proofs - which I have not yet attempted to re-interpret as a direct statement about K. Of course this purely factual criticism of certain of the results of Lambek and Szabo is in no way meant to detract from the great value of their work in demonstrating the power of proof-theory and cut-elimination techniques.

If we call a category A on which K acts a K-category, we get of course a category of K-categories, a morphism being a functor $\psi : A \to B$ commuting with the actions:

(1.1)
$$\begin{array}{ccc} K \circ A & \xrightarrow{\ \theta\ } & A \\ {\scriptstyle K\circ\psi}\downarrow & & \downarrow{\scriptstyle \psi} \\ K \circ B & \longrightarrow & B \end{array}$$

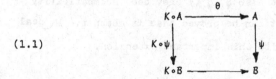

We may call these <u>strict morphisms</u> of K-categories; they preserve all the structure on the nose. However K also determines a monad on the category whose objects are functors and whose morphisms are 2-cells

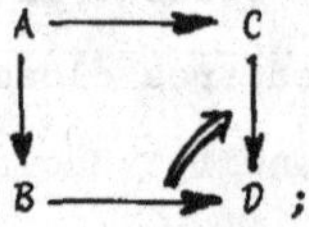

and an algebra for this monad is a $\psi: A \to B$ together with a 2-cell $\bar{\psi}$ as in

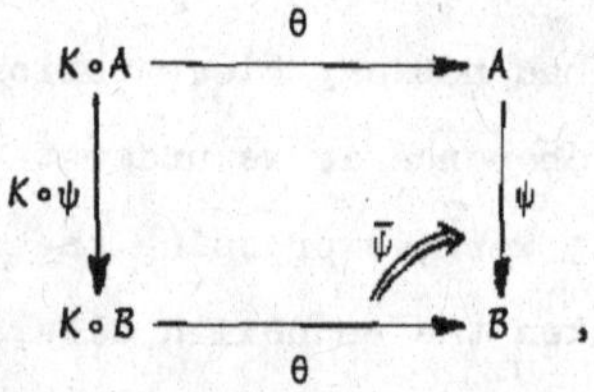

subject to the appropriate algebra-axioms. In this way we get the concept of a <u>non-strict</u> morphism $\psi, \bar{\psi}: A \to B$ of K-categories, of which a monoidal functor is the best-known example.

We proceed now to the details, but first the following sad comment: to carry out the above program in full requires a many-variable functorial calculus as in [5] for the kinds of natural transformations that occur. We therefore restrict ourselves at first to natural transformations with graphs in $\underline{\underline{P}}$ (or the corresponding $\underline{\underline{P}} \circ \Lambda \times \Lambda$ for the many-category case), for which a full calculus is given in [5]. The extension to graphs in $\underline{\underline{S}}$ or in $\underline{\underline{S}}^{op}$ is easy; graphs in $\underline{\underline{G}}$ pose yet unsolved problems; so do the mixed-variance cases. The theory can be made to work in part, however, for the mixed-variance cases corresponding to $\underline{\underline{P}}$ (and also $\underline{\underline{S}}$, $\underline{\underline{S}}^{op}$) provided incompatibility does not occur - and often it can be proved that it doesn't. We deal at some length, therefore, with this important extension.

2. Clubs and their algebras

2.1 We put ourselves in the context of §3 of the preceding paper [5], so that our category of types and graphs is $\underline{\underline{Q}} = \underline{\underline{P}}\circ\Lambda \times \Lambda$ for some indexing-set Λ, where Λ is regarded as a discrete category. We recall that we have a closed structure on $\underline{\underline{Cat}}/\underline{\underline{Q}}$, which has as a full subcategory the category $\underline{\underline{Cat}}/\Lambda$ of polycategories (= Λ- indexed families of categories). When Λ reduces to one element (the single-category case), $\underline{\underline{Cat}}/\underline{\underline{Q}}$ reduces to $\underline{\underline{Cat}}/\underline{\underline{P}}$ and $\underline{\underline{Cat}}/\Lambda$ to $\underline{\underline{Cat}}$.

We introduce the word club as a short name for "$\circ$- monoid in $\underline{\underline{Cat}}/\underline{\underline{Q}}$". A map of clubs is of course a monoid map. So a club K is a category over $\underline{\underline{Q}}$ together with functors $\mu\colon K\circ K \to K$ and $\eta\colon J \to K$ over $\underline{\underline{Q}}$, such that the following diagrams commute:

(2.1)
$$\begin{array}{ccc} K\circ K\circ K & \xrightarrow{\mu\circ K} & K\circ K \\ {\scriptstyle K\circ\mu}\downarrow & & \downarrow{\scriptstyle\mu} \\ K\circ K & \xrightarrow[\mu]{} & K \end{array}$$

(2.2)
$$\begin{array}{ccccc} J\circ K & \xrightarrow{\eta\circ K} & K\circ K & \xleftarrow{K\circ\eta} & K\circ J \\ {\scriptstyle\cong}\downarrow & & \downarrow{\scriptstyle\mu} & & \downarrow{\scriptstyle\cong} \\ K & \xrightarrow[1]{} & K & \xleftarrow[1]{} & K \end{array}$$

Since J is just Λ with the augmentation $\lambda \mapsto (1[\lambda],\lambda)$, to give η is just to give objects $\eta(\lambda)$ in K, which we denote by $\underline{\underline{1}}_\lambda$, with the appropriate augmentations. When $\Lambda = \{1\}$ this is just to give $\underline{\underline{1}} \in K$ with $\Gamma\underline{\underline{1}} = 1$. To denote the effect of μ on objects and on morphisms, we write $T(S_1, \ldots, S_n)$ for $\mu(T[S_1, \ldots, S_n])$ and $f(g_1, \ldots, g_n)$ for $\mu(f[g_1, \ldots, g_n])$. Then (2.1) becomes

(2.3) $$T(S_1, \ldots, S_n)(R_1, \ldots, R_m) = T(S_1(R_1,\ldots,R_{m_1}),\ldots,S_n(\ldots,R_m))$$

with a corresponding formula for morphisms, while (2.2) becomes

$$(2.4) \qquad T(\underline{1}_{\lambda_1}, \ldots, \underline{1}_{\lambda_n}) = T, \quad f(\underline{1}_{\lambda_1}, \ldots, \underline{1}_{\lambda_n}) = f$$

$$(2.5) \qquad \underline{1}_{\lambda}(S) = S, \quad \underline{1}_{\lambda}(g) = g,$$

where T has domain-type $n[\lambda_1, \ldots, \lambda_n]$, S has codomain-type λ, etc. As always, the name of an object serves as the name of its identity morphism.

Example 2.1 For any $A \in \underline{\underline{Cat}}/\underline{\underline{Q}}$, and in particular for any $A \in \underline{\underline{Cat}}/\Lambda$, $\{A,A\}$ is a club, as in §§2.5 and 3.4 of [5]; it is the endomorphism club of A.

Example 2.2 Taking $A = \Lambda$ in the above example we get the club $\underline{\underline{Q}} \simeq \{\Lambda,\Lambda\}$; in the case $\Lambda = \{1\}$ it reduces to $\underline{\underline{P}}$. The detailed structure will again be found in §§2.5 and 3.4 of [5].

Example 2.3 In the case $\Lambda = \{1\}$, the discrete category $\underline{\underline{N}}$ of the natural numbers ≥ 0 is a club, with $n(m_1, \ldots, m_n) = m_1 + \ldots + m_n$ as for $\underline{\underline{P}}$. Similarly, for a general Λ, $\underline{\underline{N}}_\Lambda = \underline{\underline{N}} \circ \Lambda \times \Lambda$ is a club.

Example 2.4 Any club K which is a discrete category is called a discrete club. The operation $\circ$ on $\underline{\underline{Cat}}/\underline{\underline{Q}}$ restricts to $\underline{\underline{Set}}/\underline{\underline{N}}_\Lambda$, which is also a closed category (whose internal-hom $\{-,-\}'$ differs however from the $\{-,-\}$ of $\underline{\underline{Cat}}/\underline{\underline{Q}}$!). A discrete club is just a $\circ$- monoid in this category. Thus $\underline{\underline{N}}$ and $\underline{\underline{N}}_\Lambda$ are discrete clubs.

Example 2.5 In the case $\Lambda = \{1\}$, consider a club K for which the augmentation $\Gamma: K \to \underline{\underline{P}}$ is the constant functor at $1 \in \underline{\underline{P}}$. To give η is to give $\underline{1} \in K$; to give μ is to give T(S) and f(g). The monoid-axioms say that K is a strict monoidal category with strictly-associative tensor product $T \otimes S = T(S)$ and strict identity $\underline{1}$. A typical example is the category $\underline{\underline{\Delta}}$ of natural numbers $n \geq 0$ and increasing maps $n \to m$; here $n+m$ is the tensor product and 0 is the identity $\underline{1}$.

Example 2.6 Again with $\Lambda = \{1\}$, let L be any category at all, let I be the unit category with unique object 1, and let $\Gamma(I) = 1$, $\Gamma(L) = 0$. Then $K = I + L$ is a club with $\underline{1} = 1$, $1(T) = T$, and $T(\ ) = T$ for $T \in L$.

2.2 For a club K, we have a monad $K\circ - : \underline{\underline{Cat}}/\underline{\underline{Q}} \to \underline{\underline{Cat}}/\underline{\underline{Q}}$, which restricts to $K\circ - : \underline{\underline{Cat}}/\Lambda \to \underline{\underline{Cat}}/\Lambda$. Since $\circ$ is a 2-functor ([5] §2.3 and §3.3), $K\circ -$ is actually a 2-monad on $\underline{\underline{Cat}}/\Lambda$, or on $\underline{\underline{Cat}}$ if $\Lambda = \{1\}$; and is hence a doctrine in the sense of Lawvere [11]. An algebra A for this monad on $\underline{\underline{Cat}}/\Lambda$ will be called a K-polycategory (K-category if $\Lambda = \{1\}$). The K-polycategories form a category with morphisms $\psi: A \to B$ as in (1.1).

A K-polycategory is a polycategory A with a club-map $\phi: K \to \{A,A\}$ or equivalently with an action $\theta: K\circ A \to A$; here ϕ is in $\underline{\underline{Cat}}/\underline{\underline{Q}}$ and θ in $\underline{\underline{Cat}}/\Lambda$ (or itself in $\underline{\underline{Cat}}/\underline{\underline{Q}}$ if A is not in the subcategory $\underline{\underline{Cat}}/\Lambda$). For ϕ to be a club-map is equivalent to the satisfaction by θ of the usual axioms for an action:

(2.6)
$$\begin{array}{ccc} K\circ K\circ A & \xrightarrow{\mu\circ A} & K\circ A \\ {\scriptstyle K\circ\theta}\downarrow & & \downarrow{\scriptstyle\theta} \\ K\circ A & \xrightarrow[\theta]{} & A \end{array} \qquad\qquad \begin{array}{ccc} J\circ A & \xrightarrow{\eta\circ A} & K\circ A \\ & \searrow{\scriptstyle =} & \downarrow{\scriptstyle\theta} \\ & & A \end{array}$$

Again we write $T(A_1, \ldots, A_n)$ for $\theta(T[A_1, \ldots, A_n])$ with a similar notation for morphisms; the axioms for an action can then be written like (2.3) and (2.5).

The following examples are numbered as in §2.1.

Example 2.2' Let $\Lambda = \{1\}$ and $K = \underline{P}$. Then a $\underline{P}$-category A is a strict symmetric monoidal category (essentially by Mac Lane's result in [14]).

Example 2.3' Let $\Lambda = \{1\}$ and $K = \underline{N}$. An $\underline{N}$-category A is a strict monoidal category.

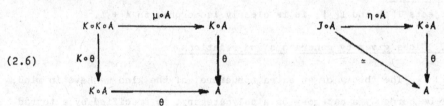

Example 2.5' Taking $\Lambda = \{1\}$ and K in Example 2.5 to be $\underline{\underline{\Delta}}$, a $\underline{\underline{\Delta}}$-category A is a category A together with a monad (T, η, μ) on A; here $TA = 1(A)$ and η, μ come from the unique maps $0 \to 1$, $2 \to 1$ in $\underline{\underline{\Delta}}$.

Example 2.6' With $\Lambda = \{1\}$ let K be as in Example 2.6. To give a K-category A is to give $L(\ ) \in A$ for $L \in L$; that is, to give a functor $L \to A$. So the category of K-categories is the category $\underline{\underline{Cat}}\backslash L$ of categories under L.

Since the free K-category on A is $K \circ A$, we have :

Example 2.2" The free strict symmetric monoidal category on A is $\underline{\underline{P}} \circ A$.

Example 2.3" The free strict monoidal category on A is $\underline{\underline{N}} \circ A$.

Example 2.5" The free category-with-a-monad on A is $\underline{\underline{\Delta}} \circ A$; since $\Gamma(\underline{\underline{\Delta}}) = 1$ this is $\Delta \times A$ (cf. [10] p.95).

Example 2.6" The free category-under-L on A is $K \circ A = (I+L) \circ A$ with objects $1[A]$ and $L[\ ]$; it is clearly isomorphic to $A + L$.

3. Clubs given by generators and relations

3.1 The theory of an extra structure, of the kind we have in mind, to be borne by a category or a polycategory, is specified by a tetrad

$$(B, \rho, D, \sigma) \tag{3.1}$$

as follows.

First there is the set B of the names of the basic functors; each of these comes with a type $\tau \in \underline{\underline{Q}}$, so that B is a set over the set $\underline{\underline{N}}_\Lambda = \underline{\underline{N}} \circ \Lambda \times \Lambda$ of objects of $\underline{\underline{Q}}$, or equally a discrete category over $\underline{\underline{Q}}$. Thus in the theory of monoidal categories, where $\Lambda = \{1\}$ and $\underline{\underline{N}}_\Lambda = \underline{\underline{N}}$, B consists of two elements $\{\otimes, I\}$ with augmentations $\Gamma\otimes = 2$ and $\Gamma I = 0$. To give a model of this much of the theory is to give a polycategory A and a functor $B \to \{A,A\}$ over $\underline{\underline{Q}}$, sending each $B \in B$ to a functor

$|B| : A_{\lambda_1} \times \ldots \times A_{\lambda_n} \to A_\mu$ of the type ΓB.

We now pass from $\mathcal{B}$ to the free discrete club $\mathcal{T}$ that it generates. We define the objects of $\mathcal{T}$ and their augmentations inductively by:

(3.2) There is an object $\underline{\underline{1}}_\lambda$ in $\mathcal{T}$ for each λ, with $\Gamma\underline{\underline{1}}_\lambda = (1[\lambda],\lambda)$.

(3.3) If $B \in \mathcal{B}$ with domain-type $\Gamma' B = n[\lambda_1, \ldots, \lambda_n]$, and if $T_1, \ldots, T_n \in \mathcal{T}$ with $\Gamma'' T_i = \lambda_i$, then there is an object $B\{T_1, \ldots, T_n\}$ in $\mathcal{T}$ with domain-type $n(\Gamma' T_1, \ldots, \Gamma' T_n)$ and codomain-type $\Gamma'' B$.

We define $\mathcal{T} \circ \mathcal{T} \to \mathcal{T}$ inductively by setting

(3.4) $\underline{\underline{1}}_\lambda(S) = S$,

(3.5) $(B\{T_1, \ldots, T_n\})(S_1,\ldots,S_m) = B\{T_1(S_1,\ldots,S_{m_1}),\ldots,T_n(\ldots,S_m)\}$,

where S, S_i are to have suitable codomain-types. Of course the unit $\mathcal{J} \to \mathcal{T}$ is given by the $\underline{\underline{1}}_\lambda$.

We identify $B \in \mathcal{B}$ with $B\{\underline{\underline{1}}_{\lambda_1}, \ldots, \underline{\underline{1}}_{\lambda_n}\} \in \mathcal{T}$. Then $B\{T_1, \ldots, T_n\}$ coincides with $B(T_1, \ldots, T_n)$ by (3.4) and (3.5), and we can now drop the curly-bracket notation in favour of round brackets.

It is clear that any functor $\mathcal{B} \to \{\mathcal{A},\mathcal{A}\}$ over $\underline{\underline{Q}}$ extends uniquely to a map $\mathcal{T} \to \{\mathcal{A},\mathcal{A}\}$ of clubs; so that a model of $\mathcal{B}$ is just a $\mathcal{T}$-algebra. Write $|T|$ for the image in $\{\mathcal{A},\mathcal{A}\}$ of $T \in \mathcal{T}$.

<u>3.2</u> The next part of the tetrad (3.1) is a relation ρ on $\mathcal{T}$, which must satisfy the condition

(3.6) $\Gamma T = \Gamma S$ if $T \rho S$.

This relation corresponds to the first lot of axioms for the structure, namely those concerned with the functors. For a model of the part $(\mathcal{B},\rho)$ of the theory is to be a model $\mathcal{A}$ of $\mathcal{B}$ in which

(3.7) $|T| = |S|$ wherever $T\rho S$.

The relation ρ is often empty in practice, and we ignored it in §1.1 above. It is non-vacuous however in the theory of strict monoidal categories; there it is given by

$$\otimes(\otimes,\underline{1})\ \rho\ \otimes(\underline{1},\otimes),$$
$$\otimes(I,\underline{1})\ \rho\ \underline{1},$$
$$\otimes(\underline{1},I)\ \rho\ \underline{1}.$$

A relation ρ (satisfying (3.6)) on any discrete club T (free or not) is called a congruence if it is an equivalence relation such that

$$T(S_1, \ldots, S_n)\ \rho\ T'(S_1', \ldots, S_n') \text{ whenever } T\rho T' \text{ and } S_i\rho S_i'.$$

Given any ρ satisfying (3.6) there is a smallest congruence $\bar{\rho}$ containing it, which can be given explicitly by the process familiar in universal algebra. The quotient $T/\bar{\rho}$ of T by this congruence is clearly again a discrete club, with its structure maps μ and η given via representatives.

Returning to case where T is the free discrete club on B constructed in §3.1, a model of (B,ρ) is a map $T \to \{A,A\}$ of clubs in which $|T| = |S|$ whenever $T\bar{\rho}S$; for the latter is clearly a consequence of (3.7). Such a model is therefore an algebra for the club $T/\bar{\rho}$. We henceforth write S for the discrete club $T/\bar{\rho}$, because the objects of S are what Kelly - Mac Lane called the shapes in [7].

3.3 The thing given by a set O of objects and a set M of morphisms or arrows with domain and codomain maps $M \rightrightarrows O$, but without composition or identities, has various names: diagram scheme, graph. The latter cannot be used here, for we have "graph" already in a different sense; I don't much care for the former; I shall call it a pre-category. Since we can speak of a map of a pre-category into a category, it makes perfectly good sense to speak of a pre-category over $\underline{Q}$. If the set $\mathcal{O}$ of objects moreover is a discrete club, I shall call $(\mathcal{O},\mathcal{M})$ a pre-club. A map of pre-clubs is of course a map of pre-categories, over $\underline{Q}$, which restricted to objects is a map of clubs.

The third part of the tetrad (3.1) is a set $\mathcal{D}$ consisting of the names of the basic natural transformations; together with domain and codomain maps $\mathcal{D} \rightrightarrows S$ making $(S,\mathcal{D})$, or $\mathcal{D}$ for short, a pre-category; provided further with a map $\Gamma\colon \mathcal{D} \to \underline{Q}$ of pre-categories, extending the augmentation Γ of S, and thus making $\mathcal{D}$ into a pre-club. A model of the part $(\mathcal{B},\rho,\mathcal{D})$ of (3.1) is a polycategory A together with a map $\mathcal{D} \to \{A,A\}$ of pre-clubs. It is therefore a model A of $(\mathcal{B},\rho)$ together with an assignment of a natural transformation $|d|\colon |T| \to |S|$ of graph ξ to each $d\colon T \to S$ in $\mathcal{D}$ with $\Gamma d = \xi$.

In the case of a (non-strict) symmetric monoidal category, for instance, the arrows of $\mathcal{D}$ are

$$a\colon \otimes(\otimes,\underline{1}) \to \otimes(\underline{1},\otimes)$$
$$\bar{a}\colon \otimes(\underline{1},\otimes) \to \otimes(\otimes,\underline{1})$$
$$r\colon \otimes(\underline{1},I) \to \underline{1}$$
$$\bar{r}\colon 1 \to \otimes(\underline{1},I)$$
$$c\colon \otimes \to \otimes,$$

all with identity graphs except c whose graph is the non-identity permutation of 2. Note that we have as yet no way of demanding that $|d|$ be a natural isomorphism; a way will be provided in §3.8 below when we introduce the second lot σ of axioms, which in the symmetric-monoidal-category case will include axioms $a\bar{a} = 1$, $\bar{a}a = 1$, $r\bar{r} = 1$, $\bar{r}r = 1$, $c^2 = 1$. In general, of course, we do not want the $|d|$ to be isomorphisms; for example, the $\tilde{\phi}$ of a monoidal functor.

We shall now construct, in several stages, the free club $\mathcal{L}$ on the pre-club $\mathcal{D}$, so that a map $\mathcal{D} \to \{A\ A\}$ of pre-clubs extends to a unique map $\mathcal{L} \to \{A,A\}$ of clubs, and a model for $(\mathcal{B},\rho,\mathcal{D})$ is just an $\mathcal{L}$-algebra.

3.4 For d: $T \to T'$ in $\mathcal{D}$ with $\Gamma d = \xi$, we define an instance of d to be a formal expression composed of d and n objects S_i of $\mathcal{D}$ (that is, of $\mathcal{S}$) of suitable codomain-types:

$$(3.8) \quad d\{S_1, \ldots, S_n\}: T(S_{\xi 1}, \ldots, S_{\xi n}) \to T'(S_1, \ldots, S_n).$$

It is assigned a domain and a codomain as in (3.8), and given the augmentation $\Gamma(d\{S_1, \ldots, S_n\}) = \xi(m_1, \ldots, m_n)$ where $\Gamma d = \xi$ and $\Gamma' S_1 = m_1[\nu_1, \ldots, \nu_{m_1}]$, etc.

For the instance $e = d\{S_1, \ldots, S_n\}$ of d and for $m = m_1 + \ldots + m_n$ objects R_i of $\mathcal{D}$ of suitable codomain-types we define

$$(3.9) \quad e(R_1,\ldots,R_m): T(S_{\xi 1} \ldots S_{\xi n})(R_{\zeta 1}\ldots R_{\zeta m}) \to T'(S_1\ldots S_n)(R_1\ldots R_m),$$

where $\zeta = \xi(m_1, \ldots, m_n)$, by setting

$$(3.10) \quad d\{S_1,\ldots,S_n\}(R_1,\ldots,R_m) = d\{S_1(R_1,\ldots,R_{m_1}),\ldots,S_n(\ldots,R_m)\}.$$

If we now identify d with its instance $d\{\mathbb{1}_{\lambda_1}, \ldots, \mathbb{1}_{\lambda_n}\}$, where $n[\lambda_1, \ldots, \lambda_n] = \Gamma' T'$, we find $d(S_1, \ldots, S_n) = d\{S_1, \ldots, S_n\}$, and we can henceforth drop curly brackets in favour of round ones.

We have now extended $\mathcal{D}$ to a bigger pre-club $\underline{\underline{\text{Inst}}}\ \mathcal{D}$, with the same objects as $\mathcal{D}$, and admitting an operation (3.9) which clearly satisfies

$$(3.11) \quad e(R_1 \ldots R_m)(V_1 \ldots V_k) = e(R_1(V_1 \ldots V_{k_1}), \ldots, R_m(\ldots V_k)),$$

$$(3.12) \quad e(\mathbb{1}_{\lambda_1}, \ldots, \mathbb{1}_{\lambda_n}) = e.$$

Any map $\mathcal{D} \to \{A,A\}$ of pre-clubs now has a unique extension to a map $\underline{\underline{\mathrm{Inst}}}\ \mathcal{D} \to \{A\ A\}$ of pre-clubs which satisfies

(3.13) $$|e(R_1, \ldots, R_m)| = |e|(|R_1|, \ldots, |R_m|).$$

<u>3.5</u> Next, for $e: S \to S'$ in $\underline{\underline{\mathrm{Inst}}}\ \mathcal{D}$, we define an <u>expansion</u> of e to be a formal expression composed of e, an object T of $\mathcal{D}$ with $\Gamma'T = n[\lambda_1, \ldots, \lambda_n]$ say, and an element $i \in n$ for which $\lambda_i = \Gamma''S = \Gamma''S'$; this formal expression is assigned a domain and a codomain, and is written as

(3.14) $$T\{\underline{\underline{1}}_{\lambda_1}, \ldots, \underline{\underline{1}}_{\lambda_{i-1}}, e, \underline{\underline{1}}_{\lambda_{i+1}}, \ldots, \underline{\underline{1}}_{\lambda_n}\}: T(\underline{\underline{1}}_{\lambda_1}, \ldots, S, \ldots, \underline{\underline{1}}_{\lambda_n}) \to T(\underline{\underline{1}}_{\lambda_1}, \ldots, S', \ldots, \underline{\underline{1}}_{\lambda_n});$$

it is given the augmentation $n(1, \ldots, 1, \eta, 1, \ldots, 1)$ where $\Gamma e = \eta$. If e is itself an instance of d, then (3.14) is called an <u>expanded instance</u> of d.

Let $f: R \to R'$ denote the expansion (3.14) of e, let P have domain-type $m[\mu_1, \ldots, \mu_m]$, let $j \in m$ with $\mu_j = \Gamma''T$, and let Q_k have codomain-type μ_k for $k \neq j$. Then we define

(3.15) $$P(Q_1, \ldots, Q_{j-1}, f, Q_{j+1}, \ldots, Q_m): P(Q_1, \ldots, R, \ldots, Q_m) \to P(Q_1, \ldots, R', \ldots, Q_m)$$

to be the expansion of e given by

(3.16) $$K\{\underline{\underline{1}}_{\mu_1}, \ldots, \underline{\underline{1}}_{\mu_{j-1}}, \underline{\underline{1}}_{\lambda_1}, \ldots, e, \ldots, \underline{\underline{1}}_{\lambda_n}, \underline{\underline{1}}_{\mu_{j+1}}, \ldots, \underline{\underline{1}}_{\mu_m}\}$$

where

(3.17) $$K = P(Q_1, \ldots, Q_{j-1}, T, Q_{j+1}, \ldots, Q_m).$$

If we now identify $e: S \to S'$, where $\Gamma''S = \Gamma''S' = \nu$, with its expansion $\underline{\underline{1}}_\nu\{e\}$, we find that $T(\underline{\underline{1}}_{\lambda_1} \ldots e \ldots \underline{\underline{1}}_{\lambda_n}) = T\{\underline{\underline{1}}_{\lambda_1} \ldots e \ldots \underline{\underline{1}}_{\lambda_n}\}$, and we can again drop curly brackets in favour of round ones.

We have now extended Inst $\mathcal{D}$ to a bigger pre-club Exp Inst $\mathcal{D}$, still with the same objects, and now admitting an operation (3.15) which clearly satisfies

$$(3.18)\ P(R_1 \ldots R_m)(V_1 \ldots f \ldots V_k)$$

$$= P(R_1(V_1 \ldots V_{k_1}), \ldots, R_i(V_p \ldots f \ldots V_q), \ldots, R_m(\ldots V_k)),$$

$$(3.19)\quad \underline{\underline{1}}_\mu(f) = f.$$

Any map Inst $\mathcal{D} \to \{A,A\}$ of pre-clubs now has a unique extension to a map Ext Inst $\mathcal{D} \to \{A,A\}$ of pre-clubs which satisfies

$$(3.20)\quad |P(Q_1, \ldots, f, \ldots, Q_m)| = |P|(|Q_1|, \ldots, |f|, \ldots |Q_m|).$$

We extend to Exp Inst $\mathcal{D}$ the operation (3.9) of §3.4 by defining

$$(3.21)\ T(\underline{\underline{1}}_{\lambda_1} \ldots e \ldots \underline{\underline{1}}_{\lambda_n})(V_1 \ldots V_k) = T(V_1, \ldots, e(V_p \ldots V_q), \ldots, V_k),$$

which easily implies the more general

$$(3.22)\quad T(R_1 \ldots f \ldots R_m)(V_1 \ldots V_k)$$

$$= T(R_1(V_1 \ldots V_{k_1}), \ldots, f(V_p \ldots V_q), \ldots, R_m(\ldots V_k)).$$

It is immediate that (3.11) and (3.12) continue to hold for $e \in$ Exp Inst $\mathcal{D}$. We conclude that any map $\mathcal{D} \to \{A,A\}$ of pre-clubs has a unique extension to a map Exp Inst $\mathcal{D} \to \{A,A\}$ of pre-clubs which satisfies both (3.13) and (3.20) for $e, f \in$ Exp Inst $\mathcal{D}$.

3.6 Write Cat Exp Inst $\mathcal{D}$ for the free category generated by the pre-category Exp Inst $\mathcal{D}$; a morphism is therefore a string $T = T_0 \to T_1 \to \ldots \to T_n = S$ of arrows of Exp Inst $\mathcal{D}$, with $n \geq 0$; composition is by concatenation, and the identities are the strings of length 0. With its evident augmentation it is again a pre-club. Any map of pre-clubs Exp Inst $\mathcal{D} \to \{A,A\}$ has a unique extension to a functor

Cat Exp Inst $\mathcal{D} \to \{A,A\}$, which is also a map of pre-clubs.

We extend to Cat Exp Inst $\mathcal{D}$ the operation (3.9) by defining $f(R_1, \ldots, R_m)$, where $f = f_k \ldots f_1$ with $f_i \in$ Exp Inst $\mathcal{D}$ and $\Gamma f_i = \eta_i$, to be

$$f_k(R_1 \ldots R_m) f_{k-1}(R_{\eta_k 1} \ldots R_{\eta_k m}) \ldots f_1(R_{\zeta 1} \ldots R_{\zeta m})$$

where $\zeta = \eta_k \eta_{k-1} \ldots \eta_2$. If $k = 0$ we of course set $1_T(R_1 \ldots R_m) = 1$. Similarly we extend the operation (3.15) by setting $P(Q_1 \ldots f \ldots Q_m)$ equal to the composite of the $P(Q_1 \ldots f_i \ldots Q_m)$; again if $k = 0$ we set $P(Q_1 \ldots 1_{Q_j} \ldots Q_m) = 1$. We still have (3.11), (3.12), (3.18), (3.19), (3.22).

We conclude that a map $\mathcal{D} \to \{A,A\}$ of pre-clubs extends to a unique map Cat Exp Inst $\mathcal{D} \to \{A,A\}$ of pre-clubs which is a functor and which satisfies (3.13) and (3.20) for $e,f \in$ Cat Exp Inst $\mathcal{D}$.

<u>3.7</u> We are now going to use the words "relation" and "congruence" in a way quite different from that of §3.2; no confusion should result.

A <u>relation</u> π on a category M is just a relation $\pi(T,S)$ on the set $M(T,S)$ for each $T, S \in M$. It is called a <u>congruence</u> if it is an equivalence relation compatible with composition. There is a smallest congruence $\bar{\pi}$ containing a relation π; $f,g\colon T \to S$ are equivalent under $\bar{\pi}$ if there is a sequence $f = f_0, f_1, \ldots, f_n = g$ of morphisms $T \to S$, where $n \geq 0$, in which each pair f_{i-1} and f_i are of the form

$$T = T_0 \to T_1 \to \ldots \to T_{k-1} \overset{u}{\underset{v}{\rightrightarrows}} T_k \to \ldots \to T_m = S$$

where either $u\pi v$ or $v\pi u$.

From a congruence π on M we get a <u>quotient category</u> M/π with the same objects as M but with equivalence classes as morphisms. When $\bar{\pi}$ is the congruence generated by the relation π, we agree to write

$\mathcal{M}/\pi$ for $\mathcal{M}/\bar{\pi}$.

Every functor $P\colon \mathcal{M} \to \mathcal{N}$ such that $Pu = Pv$ whenever $u\pi v$ factorizes uniquely through $\mathcal{M} \to \mathcal{M}/\pi$ to give a functor $\mathcal{M}/\pi \to \mathcal{N}$. If $P\colon \mathcal{M} \to \mathcal{N}$ is givenand if we <u>define</u> π by " $u\pi v \Leftrightarrow Pu = Pv$", then π is already a congruence, the <u>kernel congruence</u> of P, and $\mathcal{M}/\pi \to \mathcal{N}$ is faithful.

If $\mathcal{M}$ is augmented over $\mathbf{Q}$ the relation π is said to be <u>over</u> $\mathbf{Q}$ if $\Gamma u = \Gamma v$ whenever $u\pi v$. Then $\mathcal{M}/\pi$ has an augmentation over $\mathbf{Q}$, also called Γ.

Consider now the diagrams ($i<j$)

$$(3.23)\quad \begin{array}{ccc} T(S_1\ldots S_i\ldots S_j\ldots S_n) & \xrightarrow{T(S_1\ldots f\ldots S_j\ldots S_n)} & T(S_1\ldots S_i'\ldots S_j\ldots S_n) \\ {\scriptstyle T(S_1\ldots S_i\ldots g\ldots S_n)}\downarrow & & \downarrow{\scriptstyle T(S_1\ldots S_i'\ldots g\ldots S_n)} \\ T(S_1\ldots S_i\ldots S_j'\ldots S_n) & \xrightarrow[T(S_1\ldots f\ldots S_j'\ldots S_n)]{} & T(S_1\ldots S_i'\ldots S_j'\ldots S_n) \end{array}$$

$$(3.24)\quad \begin{array}{ccc} T(S_{\xi 1}\ldots S_{\xi\xi^{-1}i}\ldots S_{\xi n}) & \xrightarrow{h(S_1\ldots S_i\ldots S_n)} & T'(S_1\ldots S_i\ldots S_n) \\ {\scriptstyle T(S_{\xi 1}\ldots k\ldots S_{\xi n})}\downarrow & & \downarrow{\scriptstyle T'(S_1\ldots k\ldots S_n)} \\ T(S_{\xi 1}\ldots S'_{\xi\xi^{-1}i}\ldots S_{\xi n}) & \xrightarrow[h(S_1\ldots S_i'\ldots S_n)]{} & T'(S_1\ldots S_i'\ldots S_n) \end{array}$$

in **Cat Exp Inst** $\mathcal{D}$. Because $\{\mathcal{A},\mathcal{A}\}$ is a club, the images in $\{\mathcal{A},\mathcal{A}\}$ of these diagrams, under such a functor as described in the last paragraph of §3.6, commute. Such a functor, therefore, factorizes through the quotient of **Cat Exp Inst** $\mathcal{D}$ by the relation which identifies the two legs of (3.23) and the two legs of (3.24).

The same congruence on **Cat Exp Inst** $\mathcal{D}$ will clearly result if we identify the legs of (3.23) and those of (3.24) only in the special case f,g,h,k ∈ Exp Inst $\mathcal{D}$. We can go further; we can suppose f,g,k to be in Inst $\mathcal{D}$, for the functors that "expand" them can be absorbed in T or T'. Since (3.23) and (3.24) seem to assert respectively the "functoriality of T" and the "naturality of h", we call them "relations of functoriality" and "relations of naturality".

Write, then, Funct for the relation on Cat Exp Inst $\mathcal{D}$ which identifies the legs of (3.23) for f,g ∈ Inst $\mathcal{D}$, and Nat for the relation which identifies the legs of (3.24) for k ∈ Inst $\mathcal{D}$, h ∈ Exp Inst $\mathcal{D}$; write Funct + Nat for the union of these two relations. Define

(3.25) $\mathcal{L}$ = Cat Exp Inst $\mathcal{D}$/(Funct + Nat).

Now $\mathcal{L}$ admits the structure of a club. Perhaps the easiest way to see this is to observe that the operations (3.9) and (3.15) on Cat Exp Inst $\mathcal{D}$ are compatible with (3.23) and (3.24) and hence carry over to $\mathcal{L}$. Then (3.23) and (3.24) can be written in $\mathcal{L}$ and commute. The diagonal of (3.23) gives an operation $T(S_1 \dots f \dots g \dots S_n)$, and by extension an operation $T(k_1, \dots, k_n)$. The analogue of (3.24) with the right vertical arrow replaced by $T'(k_1, \dots, k_n)$ still commutes, and the diagonal of this diagram gives the operation $h(k_1, \dots, k_n)$. The map $h[k_1, \dots, k_n] \mapsto h(k_1, \dots, k_n)$ is clearly functorial, and the club axioms follow easily from (3.11), (3.12), (3.18), (3.19), (3.22).

The map $\mathcal{L} \to \{A,A\}$ induced by the functor Cat Exp Inst $\mathcal{D} \to \{A,A\}$ at the end of §3.6 is clearly a map of clubs; and we can assert that a map $\mathcal{D} \to \{A,A\}$ of pre-clubs extends to a unique map of clubs $\mathcal{L} \to \{A,A\}$. (The word "extends" is legitimate: since the legs of (3.23) and (3.24) are of length 2 in Cat Exp Inst $\mathcal{D}$, no two elements of Exp Inst $\mathcal{D}$ are identified in $\mathcal{L}$; hence Exp Inst $\mathcal{D}$, and *a fortiori* $\mathcal{D}$ itself, are faithfully embedded in $\mathcal{L}$.)

We conclude that a model for the part $(\mathcal{B},\rho,\mathcal{D})$ of (3.1) is the same thing as a polycategory A with a club-map $\mathcal{L} \to \{A,A\}$; that is, an $\mathcal{L}$-algebra or an $\mathcal{L}$-polycategory.

3.8 The last part of the tetrad (3.1) is a relation σ over $\mathbf{Q}$ on $\mathcal{L}$; a model of $(\mathcal{B},\rho,\mathcal{D},\sigma)$ is now a model of $(\mathcal{B},\rho,\mathcal{D})$ - that is, an $\mathcal{L}$-algebra A - for which $|f| = |g|$ whenever $f\sigma g$.

More commonly, σ is given as a relation not on $\mathcal{L}$ but on **Cat Exp Inst** $\mathcal{D}$; it makes no difference since a relation σ on the latter determines a relation $[\sigma]$ on $\mathcal{L}$ where $[f]$ and $[g]$ are related under $[\sigma]$ if <u>some</u> representatives f and g of them are related under σ, and since any relation on $\mathcal{L}$ so arises.

For example, the relation σ on **Cat Exp Inst** $\mathcal{D}$ for a symmetric monoidal category, where $\mathcal{D}$ is as in §3.3, is given as follows (cf.[7] page 98):

$$
\begin{aligned}
a(\underline{1},\underline{1},\otimes)\cdot a(\otimes,\underline{1},\underline{1}) \;&\sigma\; \otimes(\underline{1},a)\cdot a(\underline{1},\otimes,\underline{1})\cdot \otimes(a,\underline{1})\\
\otimes(\underline{1},r)\cdot a(\underline{1},\underline{1},I) \;&\sigma\; r(\otimes)\\
c\cdot c \;&\sigma\; 1\\
a\cdot c(\otimes,\underline{1})\cdot a \;&\sigma\; \otimes(\underline{1},c)\cdot a\cdot \otimes(c,\underline{1})\\
a\cdot\bar{a} \;&\sigma\; 1\\
\bar{a}\cdot a \;&\sigma\; 1\\
r\cdot\bar{r} \;&\sigma\; 1\\
\bar{r}\cdot r \;&\sigma\; 1
\end{aligned}
$$

The last four of these are our way of guaranteeing that $|a|$, $|r|$ are natural <u>isomorphisms</u> in $\{A,A\}$; the third already guarantees this for $|c|$.

If $\mathcal{L}$ any club, a <u>club-congruence</u> on $\mathcal{L}$ is a congruence π over $\mathbf{Q}$ on $\mathcal{L}$, in the sense of §3.7, which further satisfies

(3.26) $f\pi f'$ and $g_1 \pi g_1'$ imply $f(g_1, \ldots, g_n)\ \pi\ f'(g_1', \ldots, g_n')$.

If π is a club-congruence, $\mathcal{L}/\pi$ is again a club, called a <u>quotient club</u>

of L (not to be confused with the meaning of "quotient" for discrete clubs in §3.2, to which we shall seldom have to refer). The kernel-congruence π of a map $L \to M$ of clubs is automatically a club-congruence, and we get club-maps $L \to L/\pi \to M$; we call L/π the image of $L \to M$.

Any congruence π on L is contained in a smallest club-congruence $\hat{\pi}$. If π is generated by a relation σ, define a new relation τ by

(3.27) $M(L_1 \ldots f(N_1 \ldots N_k) \ldots L_n) \ \tau \ M(L_1 \ldots f'(N_1 \ldots N_k) \ldots L_n)$ if $f \sigma f'$;

so that τ consists of the "expanded instances of σ". Then the congruence $\bar{\tau}$ generated by τ is clearly the club-congruence $\hat{\pi}$ generated by $\bar{\sigma} = \pi$.

Returning now to the club L of (3.25), an L-algebra A for which $|f| = |g|$ whenever $f \sigma g$ is the same thing as a K-algebra, where K is the club

(3.28) $$K = L/\tau$$

and τ is given by (3.27). This then is what a model of $(\mathcal{B},\rho,\mathcal{D},\sigma)$ is.

Let us now translate to the more usual case where σ is given as a relation over Q not on L but on Cat Exp Inst $\mathcal{D}$. Again define τ by (3.27) to consist of the expanded instances of σ; now τ is a relation on Cat Exp Inst $\mathcal{D}$. Then the image $[\tau]$ of τ in L consists precisely of the expanded instances of the image $[\sigma]$ of σ. The club K of (3.28), which must now be written as $L/[\tau]$, is therefore Cat Exp Inst $\mathcal{D}$/(Funct + Nat + τ). We give a new name Imp to this relation τ on Cat Exp Inst $\mathcal{D}$, to suggest the (expanded instances of the) imposed relations. Our final result, then, is that a model of $(\mathcal{B},\rho,\mathcal{D},\sigma)$ is precisely a K-algebra for the club

(3.29) $K = \underline{\underline{Cat}}\ \underline{\underline{Exp}}\ \underline{\underline{Inst}}\ \mathcal{D}/(\underline{\underline{Funct}} + \underline{\underline{Nat}} + \underline{\underline{Imp}})$.

We sum up our conclusion as:

Theorem 3.1 The data $(\mathcal{B},\rho,\mathcal{D},\sigma)$ describing an extra structure for polycategories A determine a club K such that a model for $(\mathcal{B},\rho,\mathcal{D},\sigma)$ is precisely a K-polycategory A. The discrete club S of objects of K is the quotient of the free discrete club on $\mathcal{B}$ by the congruence generated by ρ. As a category, K is generated by the expanded instances of the elements of $\mathcal{D}$, subject to the functoriality and naturality relations (3.23) and (3.24), and to the expanded instances of the imposed relations σ. Its structure as a club is derived from the operations (3.9) and (3.15).

3.9 The fact that the only relations are Funct + Nat + Imp is the "important theoretical conclusion" we referred to in the second paragraph of §1.4; a diagram in Cat Exp Inst $\mathcal{D}$ commutes in K if and only if it can be filled in in the classical way with a trellis of little diagrams of these kinds. A simple application of this is the following.

Suppose we have two theories $(\mathcal{B},\mathcal{D},\sigma)$ and $(\mathcal{B}',\mathcal{D}',\sigma')$ where $\mathcal{B}' \supset \mathcal{B}$, $\mathcal{D}' \supset \mathcal{D}$, and $\sigma' \supset \sigma$; for simplicity we take ρ,ρ' to be vacuous. Then $S' \supset S$, and Cat Exp Inst $\mathcal{D}' \supset$ Cat Exp Inst $\mathcal{D}$. We get a map of clubs $K \to K'$.

Suppose that for every $d \in \mathcal{D}'-\mathcal{D}$, either the domain or the codomain of d lies in $S'-S$. Then if either f or g in (3.23) is an expanded instance of d, neither leg of (3.23) lies in Cat Exp Inst $\mathcal{D}$; similarly if either h or k in (3.24) is an expanded instance of d, neither leg lies in Cat Exp Inst $\mathcal{D}$.

Now suppose further that, whenever u, v: P → Q are related by σ' but not by σ, neither u nor v lies in Cat Exp Inst $\mathcal{D}$. It then easily follows that if f, g: T → S in Cat Exp Inst $\mathcal{D}$ are congruent under Funct' + Nat' + Imp', they are already congruent under Funct + Nat + Imp; in other words, $K \to K'$ is faithful. The commutativity of any diagram in K can then be deduced from the commutativity of its image in K'.

4. Extensions to more general natural transformations

4.1 The extension of the above to the case where the functors are still covariant but the natural transformations have graphs in $\underline{S}$ or in $\underline{S}^{op}$ (as described in §4.1 of [5]) would seem to be quite straightforward, although I haven't checked the details. $\underline{S}^{op}$ itself is the club for the extra structure "posessing strictly-associative finite products", and $\underline{S}^{op} \circ A$ is the free such category on A; an object of this is $n[A_1, \ldots, A_n]$ where $A_i \in A$, and a morphism $n[A_1, \ldots, A_n] \to m[B_1, \ldots, B_m]$ consists of a function ψ: m → n and maps $A_{\psi i} \to B_i$ in A. If we generalize a bit further as suggested in the last paragraph of §4.1 of [5], we can similarly describe the free category-with--equalizers on A. The more general covariant case with graphs in $\underline{G}$ cannot be handled in these terms until a calculus like that of [5] is developed for this case.

4.2 A more hopeful case is that of mixed-variance functors and natural transformations of the every-variable-twice kind introduced by Eilenberg-Kelly in [1]. We refer the reader to §4.2 of [5] for a discussion of the obstacles in the way of a suitable calculus. Presumably if the right category $\underline{P}^*$ of graphs-with-closed-loops can be found as discussed there, everything above will carry over at once. In the meantime we must do what we can with the category $\underline{P}^*_o$ of simple graphs (those without closed loops). We shall deal here for simplicity with the one-category case, the extension to polycategories presenting no difficulties at all. We assume familiarity both with

[1] and with §4.2 of [5]; $\mathbf{P}^*_o$ is defined in the latter.

We call a category K over $\mathbf{P}^*_o$ simple if, for any composable morphisms $f: T \to S$ and $g: S \to R$ of K, the graphs Γf and Γg are compatible. We write $\mathbf{Sim}/\mathbf{P}^*_o$ for the category of simple categories over $\mathbf{P}^*_o$ and all functors between them; more properly it is a 2-category. It contains $\mathbf{Cat}$ as a full subcategory, when $A \in \mathbf{Cat}$ is given as augmentation the constant functor at $0[\ ]$. (Recall that an object of $\mathbf{P}^*_o$ is $n[\sigma_1, \ldots, \sigma_n]$ where each σ_i is + or -.)

We indicated in §4.2 of [5] how to define an associative operation $A \circ B$ in $\mathbf{Sim}/\mathbf{P}^*_o$, with values in $\mathbf{Cat}$ if $B \in \mathbf{Cat}$. Objects are like $A[B_1, \ldots, B_n]$, where $\Gamma A = n[\sigma_1, \ldots, \sigma_n]$ with $\sigma_i = \pm$; a morphism $A[B_1, \ldots, B_n] \to A'[B_{n+1}, \ldots, B_{n+m}]$ is $f[g_1, \ldots, g_k]$ where $f: A \to A'$ with $\Gamma f = \xi$, where $2k = n+m$, and where $g_i: B_p \to B_q$; here $p, q \in n + m$ are mates under ξ such that p has the - sign and q the + sign in $(-\sigma_1, -\sigma_2, \ldots, -\sigma_n, \sigma_{n+1}, \ldots, \sigma_{n+m})$. As for the order in which $g_1, \ldots, g_k$ are written, we may as well take it to be the order of their codomain indices in $n + m$. Because of the simplicity, composition presents no problems; we wind happily around the graphs composing as we go:

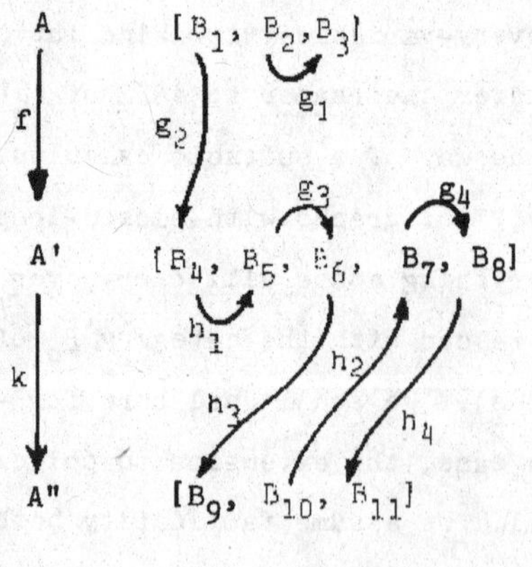

We augment $A \circ B$ over $\underline{\underline{P}}_0^*$, when B has a non-trivial augmentation, by the obvious process of replacing linkages by entire graphs; the identity J for $\circ$ is the unit category I with augmentation $1[+]$. The one thing we don't have is a functor category $\{B,C\}$ right adjoint to $A \circ B$, for we have no way of composing natural transformations of incompatible graphs.

We can still define a <u>club</u> K in $\underline{\underline{Sim}}/\underline{\underline{P}}_0^*$, namely as a $\circ$- monoid. We can also define a K-algebra, or K-category; not, it is true, by a club-map $K \to \{A,A\}$, but by what is equivalent to this in the covariant case, namely an <u>action</u> $K \circ A \to A$.

Now suppose we are given data (B,ρ,D,σ) as in (3.1) to describe an extra structure on a category A. Now B is a set over the set $\underline{\underline{N}} \circ 2$ of objects of $\underline{\underline{P}}_0^*$, and ρ is again a relation respecting the augmentation. Always modulo the replacement of $M \to \{A,A\}$ by $M \circ A \to A$ etc., we can take over §§3.1 and 3.2 unchanged. We must suppose of D of course that it is augmented over $\underline{\underline{P}}_0^*$. We then meet no problems in carrying over §3.3 - §3.5. When we come to §3.6, it may happen that there are f: T → S, g: S → R in $\underline{\underline{Exp}}$ $\underline{\underline{Inst}}$ D with incompatible graphs; if so we are stymied, and we merely say (with apologies to Dr.Johnson) that (B,ρ,D,σ) is <u>not clubbable</u>. This may happen even though the axioms σ require us to compose no incompatibles; an example is the theory of a <u>compact closed category</u> as given in §4.2 of [5]. On the other hand, composables in $\underline{\underline{Exp}}$ $\underline{\underline{Inst}}$ D may always be compatible; this has been shown to be the case in [7], [8], and [12] for the theories of closed categories, of natural transformations enriched over a closed category, and of maps of closed categories; a later paper [6] by the author in this volume will show it to be the case whenever the only contravariance arises through positing adjoints to covariant functors. In these good cases where all composables are compatible, we may carry over §3.6 - §3.8, obtaining the analogue of Theorem 3.1; such a case is said to be clubbable. Of course §3.9 too carries over.

To extend to the many-category case is easy: just replace $\underline{\underline{P}}_0^*$ as category of types and graphs by the appropriate $\underline{\underline{Q}}_0^*$, which depends on Λ. A type is now $(n[\pm\lambda_1, \ldots, \pm\lambda_n], \mu)$ and a graph is as before except that it must mate elements corresponding to the same λ.

4.3 Let us return to the unclubbable case. To say that we are stymied does not mean that we are out of the game - only that we cannot describe a model of $(\mathcal{B},\rho,\mathcal{D},\sigma)$ as an algebra for some club K. We successfully carried over §§3.1-3.2 to get the discrete club S of objects, and further - taking $\mathcal{D}$ to be augmented over $\underline{\underline{P}}_0^*$ - carried over §§3.3-3.5 to get the pre-club $\underline{\underline{\text{Exp}}}\ \underline{\underline{\text{Inst}}}\ \mathcal{D}$. When we come to §3.6, we can still form the free category $\underline{\underline{\text{Cat}}}\ \underline{\underline{\text{Exp}}}\ \underline{\underline{\text{Inst}}}\ \mathcal{D}$ on this, and augment it over $\underline{\underline{P}}_0^*$; it is just not a simple category in the sense of §4.1. We can still write the analogues of (3.23) and (3.24) in §3.7 for $f, g, k \in \underline{\underline{\text{Inst}}}\ \mathcal{D}$ and $h \in \underline{\underline{\text{Exp}}}\ \underline{\underline{\text{Inst}}}\ \mathcal{D}$, and form the quotient category L as in (3.25); again it would have made no difference if we had taken $f, g, k \in \underline{\underline{\text{Exp}}}\ \underline{\underline{\text{Inst}}}\ \mathcal{D}$. Finally, coming to §3.8, suppose that the relation σ on $\underline{\underline{\text{Cat}}}\ \underline{\underline{\text{Exp}}}\ \underline{\underline{\text{Inst}}}\ \mathcal{D}$ only relates pairs u,v which, written as morphisms $u_m \ldots u_2 u_1$ and $v_n \ldots v_2 v_1$ with $u_i, v_j \in \underline{\underline{\text{Exp}}}\ \underline{\underline{\text{Inst}}}\ \mathcal{D}$, have the u_i compatible and the v_j compatible (see [7], Lemma 5.3, for this concept). Then the definition of $\tau = \underline{\underline{\text{Imp}}}$ in (3.27) continues to make excellent sense, and we can form the category K as in (3.29).

Since K is not simple, we cannot form $K \circ A$, which would in the clubbable case be the free model on A. But in that case the free model $K \circ \Lambda$ on Λ is equal to K as a category, apart from its augmentation; and even in the unclubbable case we still have this:

Theorem 4.1 The category K given by (3.29) is the free model of $(\mathcal{B},\rho,\mathcal{D},\sigma)$ on the category Λ.

<u>Proof</u> It <u>is</u> a model: first of all a model of $(\mathcal{B},\rho)$, since for $T \in \mathcal{S}$ we get a functor $|T|: \mathcal{K}^{\nu} \to \mathcal{K}$ by setting $|T|(S_1, \ldots, S_n) = T(S_1, \ldots, S_n)$ and $|T|(g_1, \ldots, g_n) = T(g_1,\ldots,g_n)$; the latter is defined in the first instance for $g_i \in$ **Exp Inst** $\mathcal{D}$, using (3.23), and for arbitrary g_i by composition. Next it is a model of $(\mathcal{B},\rho,\mathcal{D})$, since for $d \in \mathcal{D}$ we get a natural transformation $|d|$ with components $|d|(S_1, \ldots, S_k) = d(S_1, \ldots, S_k)$; it is natural by the obvious extension of (3.24). Finally it is a model of $(\mathcal{B},\rho,\mathcal{D},\sigma)$, since $u\ \sigma\ v$ implies $u(S_1, \ldots, S_n)\ \tau\ v(S_1, \ldots, S_n)$ by (3.27), whence $|u| = |v|$.

If $\mathcal{A}$ is <u>any</u> model, a functor $\Lambda \to \mathcal{A}$ is in effect a family of objects $K_\lambda \in \mathcal{A}$. We are to show that there is a unique map $\phi: \mathcal{K} \to \mathcal{A}$ of models with $\phi(\underline{1}_\lambda) = K_\lambda$. Certainly there is a unique such map $\phi: \mathcal{S} \to \mathcal{A}$ of models of $(\mathcal{B},\rho)$, since $(\mathcal{B},\rho)$ alone is certainly clubbable (or directly from (3.2) and (3.3)). The value of ϕ is then forced on **Exp Inst** $\mathcal{D}$, extends to **Cat Exp Inst** $\mathcal{D}$, and finally factorizes through $\mathcal{K}$ since it respects **Funct**, **Nat**, and **Imp**. □

It is valuable to be able to talk about $\mathcal{K}$ whether it is a club or not, not so much because the free model on Λ is of great interest in itself, but because in practice, even when $\mathcal{K}$ <u>is</u> a club, we may be able to prove this only <u>a posteriori</u>, perhaps by a cut-elimination argument, without having proved it directly at the §3.6-stage of the construction; we do precisely this in the paper [6] in this volume.

Although we cannot in the unclubbable case get the free model on $\mathcal{A}$ as $\mathcal{K} \circ \mathcal{A}$, which is now undefined, we can still construct directly the free model on $\mathcal{A}$, much as we constructed $\mathcal{K}$. In fact the simplest way to express it is this: form from $(\mathcal{B},\rho,\mathcal{D},\sigma)$ a new theory $(\mathcal{B}',\rho',\mathcal{D}',\sigma')$ by adding to $\mathcal{B}$ the objects A of $\mathcal{A}$, with augmentation $A \mapsto (0[\ ], \Gamma A)$, to get $\mathcal{B}'$; by setting $\rho' = \rho$; by adding to $\mathcal{D}$ the morphisms of A to get $\mathcal{D}'$; and by adding to σ the relations $h = fg$ and $k = 1_A$, where h is the actual composite fg in A and k is id_A in A, to get σ'. Form the K' for this theory, and let $K[A]$ be the full subcategory of K' determined by the objects of domain-type $0[\ ]$. It is then easy to see that $K[A]$ is the free $(\mathcal{B},\rho,\mathcal{D},\sigma)$ - model on A; its set of objects is of course $S \circ \underline{\underline{Ob}}A$.

The monad $K[\ ]$ is just not determined by knowledge of K as a category over $\underline{\underline{Q}}^*_o$; we lost too much information by throwing away the closed loops when we composed incompatibles in $\underline{\underline{Q}}^*_o$. If we take the example of <u>compact closed categories</u> given in §4.2 of [5], then the morphism (4.7) there in the corresponding K has empty graph in $\underline{\underline{P}}^*_o$, and fails to inform us that there will be morphisms like (4.8) there in $K[A]$ for $u,v \in A$. Once again it is clear that a suitable $\underline{\underline{P}}^*$ if found would help; (4.7) would then have a graph consisting of a single closed loop, $K \circ A$ would now make sense, and would contain a morphism (4.8) depending on a cycle in A. The conviction is strengthened that $K[A]$ is of the form $K \circ A$ for a monoid K in $\underline{\underline{Cat}}/\underline{\underline{P}}^*$ for some suitable, but not-yet-found, $\underline{\underline{P}}^*$.

<u>4.4</u> The ideas in §4.2 above surely extend to the mixed-variance cases corresponding to graphs in $\underline{\underline{S}}$ or in $\underline{\underline{S}}^{op}$. A graph in $\underline{\underline{S}}^{op*}_o$ from $n[\sigma_1, \ldots, \sigma_n]$ to $m[\sigma_{n+1}, \ldots, \sigma_{n+m}]$, where σ_i is + or −, is now a <u>function</u>, and not a <u>bijection</u>, from the +'s to the −'s in $(-\sigma_1, \ldots, -\sigma_n, \sigma_{n+1}, \ldots, \sigma_{n+m})$. There is a simple concept of compatibility for such graphs - namely the absence of closed paths in the composite graph - and there is an evident way of composing compatible graphs. I don't see how to make $\underline{\underline{S}}^{op*}_o$ a category: one can no longer define a composite of incompatibles just by "discarding the closed loops"; but in the clubbable case, where incompatibles just don't occur, §4.2 surely carries over. The theory of <u>cartesian closed categories</u> has natural transformations of this kind. It will

certainly be clubbable by Szabo's cut-elimination theorem ([15],[16]) for this theory. I rather suspect that what Szabo calls the "Generality Theorem" in [16] and the "Coherence Theorem for CCM" in [15] will turn out to be the assertion that $\Gamma: K \to \underline{S}^{op^*}_{o}$ is faithful; in other words that "all diagrams commute".

5. Coherence - comparison with Kelly-Mac Lane

5.1 The coherence problem for $(B,\rho,\mathcal{D},\sigma)$, in the weakest sense, consists in determining which diagrams Δ in $\underline{\underline{Cat}}\ \underline{\underline{Exp}}\ \underline{\underline{Inst}}\ \mathcal{D}$ have commutative images under the functor

$$(5.1)\qquad \psi: \underline{\underline{Cat}}\ \underline{\underline{Exp}}\ \underline{\underline{Inst}}\ \mathcal{D} \to \underline{\underline{Cat}}\ \underline{\underline{Exp}}\ \underline{\underline{Inst}}\ \mathcal{D}/(\underline{\underline{Funct}} + \underline{\underline{Nat}} + \underline{\underline{Imp}}) = K;$$

in the strongest sense, however, it consists in the complete determination of the club K, with its augmentation, and with its club-structure, from the generators and relations $(B,\rho,\mathcal{D},\sigma)$; we then of course know the free K-polycategory $K\circ A$ on any A, including its structure as a K-algebra. (We are restricting ourselves entirely to the clubbable case.)

To relate this to the formulation used for instance by Kelly-Mac Lane in [7], which talked entirely in terms of models, it is convenient to introduce a canonical factorization of club-maps.

A map $\alpha: K \to L$ of clubs has a kernel-congruence π which is a club-congruence, and α factorizes as $K \to K/\pi \to L$; as we said in §3.8 we call K/π the image of α, written im α; the club map $K/\pi \to L$ is faithful, while $K \to K/\pi$ is a quotient map of clubs. Recall that K/π has the same objects as K. Introduce a new club wim α called the wide image of α; its objects are to be those of K, but a morphism $T \to S$ is to be a morphism $\alpha T \to \alpha S$ in L. Its club-structure is obvious, and $\alpha: K \to L$ now factorizes into club-maps

$$(5.2)\qquad K \to K/\pi = \text{im}\,\alpha \to \text{wim}\,\alpha \to L;$$

the first is a quotient map, the second is faithful and (like the first) is the identity on objects; the third is faithful and "the identity" on morphisms. We may identify the morphisms of $K/\pi = \mathrm{im}\,\alpha$ with their images in wim α. Clearly the factorization (5.2) is functorial.

In the covariant case, let A be a K-algebra, and apply the factorization (5.2) to ϕ and to $\Gamma\colon K \to \mathbf{Q}$ in the commutative diagram

$$
(5.3) \qquad
\begin{array}{ccc}
K & \xrightarrow{\ \phi\ } & \{A,A\} \\
 & \searrow^{\Gamma} & \downarrow{\scriptstyle \Gamma} \\
 & & \mathbf{Q}
\end{array}
$$

to get a diagram we shall write as

$$
(5.4) \qquad
\begin{array}{ccccccc}
K & \xrightarrow{\ \phi_1\ } & \mathbf{All}\ N(A) & \xrightarrow{\ \phi_2\ } & N(A) & \longrightarrow & \{A,A\} \\
 & \searrow^{\Gamma_3} & \downarrow{\scriptstyle \Gamma_2} & & \downarrow{\scriptstyle \Gamma_1} & & \downarrow{\scriptstyle \Gamma} \\
 & & \mathbf{All}\ G & \longrightarrow & G & \longrightarrow & \mathbf{Q}.
\end{array}
$$

Note further that if $\alpha\colon K \to L$ is a club-map, L becomes a K-algebra via $K \circ L \to L \circ L \to L$; so all the objects in (5.4) are not only clubs but K-algebras, while all the maps are not only club-maps but maps of K-algebras.

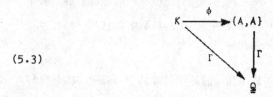

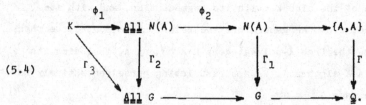

Passing now to the mixed-variance (but clubbable) case, we cannot copy all of this, since we have no $\{A,A\}$ and since the $\mathbf{Q}^*_0$ corresponding to $\mathbf{Q}$ is not a club. We can however write $N(A)$ (supposing A to be non-empty) by adopting some harmless definition of composition for incompatible natural transformations: pick some object of A and agree to compose by setting all arguments occurring in closed loops equal to this object. We can write the bottom line since $\mathbf{Q}^*_0$ is at any rate a category; we compose by ignoring closed loops. We can therefore write

$$
(5.5)\qquad
\begin{array}{ccccc}
K & \xrightarrow{\ \phi_1\ } & \mathrm{All}\ N(A) & \xrightarrow{\ \phi_2\ } & N(A) \\
 & \searrow^{\Gamma_3} & \downarrow{\scriptstyle \Gamma_2} & & \downarrow{\scriptstyle \Gamma_1} \\
 & & \mathrm{All}\ G & \longrightarrow & G\ .
\end{array}
$$

Here All $N(A)$ and All G are clearly clubs, namely quotient clubs of K. The other categories $N(A)$ and G, while not clubs since not simple, are at any rate K-algebras; and all the maps are K-algebra maps.

All $N(A)$ is what, in §1.3 of the Introduction, we called All(A), and K is what we called the "universal All". We have changed notation now to allow comparison with the formulation of Kelly-Mac Lane in §2 of [7].

G is what they called G, the "category of shapes and graphs"; $N(A)$ is what they called N = N(A), the "category of shapes and natural transformations" (they had V for A). These were K-algebras, that is, closed categories in the case of [7]. All $N(A)$ and All G are what they called the subcategories of <u>allowable morphisms</u> in $N(A)$ and in G respectively; they were defined there as the smallest sub-K-categories containing all the objects, which clearly agrees with their present definition.

The results of [7] were (i) that there was an algorithm for finding the maps $T \to S$ in All G; (ii) that any two such maps were compatible, so that the case is indeed a clubbable one; (iii) that Γ_2 was surjective, trivial for us because Γ_3 is a quotient map; and finally (iv) that Γ_2, when restricted to the full subcategory of All $N(A)$ determined by a certain subset Prop S of its set of objects S, was faithful.

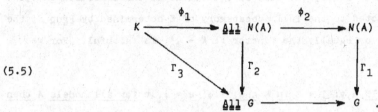

In our terms, the last is equivalent to the assertion that, when restricted to the full subcategory of K determined by $\underline{\underline{Prop}}\ S$, the functor Γ_3, or equally the functor $\Gamma: K \to \underline{\underline{Q}}_o^*$, is faithful. For we have :

Lemma 5.1 *If* $u, v: T \to S$ *in* K *and if* $\phi_1\ u = \phi_1\ v$ *for all models* A *then* $u = v$.

Proof Take the model A to be K itself, seen as a K-algebra via the map $\mu: K \circ K \to K$. The $(R_1, \ldots, R_n)$ - component of $\phi_1 u = |u|$ is then precisely $u(R_1, \ldots, R_n)$; the $(\underline{1}, \ldots, \underline{1})$ - component is therefore u itself; so $\phi_1\ u = \phi_1\ v$ implies $u = v$. □

The club $\underline{\underline{All}}\ N(A)$ is a quotient club of K intermediate between K and $\underline{\underline{All}}\ G$; we have just seen that it is equal to K when $A = K$; it reduces to $\underline{\underline{All}}\ G$ when A is the unit category I (which is always a model). From our present point of view it is a red herring; the proper objects of interest are K, and its image $\underline{\underline{All}}\ G$ under $\Gamma: K \to \underline{\underline{Q}}_o^*$.

5.2 The complete solution of the coherence problem requires first of all the determination of the set S of objects of K from the data (B, ρ). This presents of course no difficulty if ρ is vacuous, but otherwise leads to a word-problem which may be anything from trivial to insoluble.

The next step might often be in practice the determination, perhaps as in [7] via an algorithm, of the image $\underline{\underline{All}}\ G$ of $\Gamma: K \to \underline{\underline{Q}}_o^*$. Then it remains to determine for each $\xi: T \to S$ in $\underline{\underline{All}}\ G$ the inverse image $\Gamma^{-1}\xi$; to know K as a category we must then determine its law of composition; and finally to know it as a club we must determine its law of substitution $f(g_1, \ldots, g_n)$.

In one case these last two steps are unnecessary, namely when

Γ is faithful. In this "all diagrams commute" case, the job is finished when we know $\underline{\underline{All}}\ G$, for we know how to compose and to substitute graphs, and this forces composition and substitution in K. The typical example of this case is Mac Lane's result [14] for symmetric monoidal categories; there the category of graphs is $\underline{\underline{P}}$, and $\underline{\underline{All}}\ G$ is all of $\underline{\underline{P}}$, so that $\Gamma: K \to \underline{\underline{P}}$ is an equivalence of categories; the only difference between K and $\underline{\underline{P}}$ is that K has more objects. For strict symmetric monoidal categories we have of course $K = \underline{\underline{P}}$.

5.3 We do not mean to suggest that all interesting extra structures on a category or a polycategory are given by data and axioms $(B,\rho,\mathcal{D},\sigma)$ as in §3. Certainly there may be axioms not of the form ρ and σ; typically we may have a theory

$$(5.6) \qquad (B,\rho,\mathcal{D},\sigma;\tau),$$

a model of which is a model of $(B,\rho,\mathcal{D},\sigma)$ satisfying further axioms τ not writable in the forms ρ and σ. I have not yet come across such a theory in nature, however, for which the coherence problem differs from that of $(B,\rho,\mathcal{D},\sigma)$.

One example is the problem of a natural transformation enriched over a closed category, studied by Kelly-Mac Lane [8] in this volume. This is a 3-category problem; and if we write A_1, A_2, A_3 for the $\underline{\underline{A}}$, $\underline{\underline{A}}'$, $\underline{\underline{V}}$ of [8], the theory is of the form (5.6) where τ is the assertion that A_1 and A_2 are discrete. But the club K for the part $(B,\rho,\mathcal{D},\sigma)$ of the theory already has K_1 and K_2 discrete (where K_i is the part of K with codomain-type i), and there is no coherence problem other than that for all K-algebras.

Again, consider the theory of "closed categories without $\otimes$", as given in Chapter I §2 of [2]. First let us change the theory inessentially, to bring it closer to our present form, by dropping what was there called the functor V, which is in any case representable by I. Then $\mathcal{B}$ consists of two elements [,] (replacing the (,) of [2]) and I, of appropriate types; ρ is vacuous; $\mathcal{D}$ consists of L, j, i as there, together with $\bar{\imath}$ to act as the inverse of i; σ consists of the axioms CC1 - CC4 of [2], together with $i\bar{\imath} = 1$ and $\bar{\imath}i = 1$, and finally $j_I = i_I$. Then a closed-category-without-⊗ as in [2] is essentially a model A of $(\mathcal{B},\rho,\mathcal{D},\sigma)$ satisfying the extra axiom τ:

(5.7) The map κ: $A(A,B) \to A(I,[A,B])$, given by $f \longmapsto [1,f]j_A$, is an isomorphism.

In the coherence problem for such categories, we may want to prove that $u = v: A \to B$ in A, while the solution to the coherence problem for $(\mathcal{B},\rho,\mathcal{D},\sigma)$ may only give us $\kappa u = \kappa v$; but then (5.7) gives what we want. I haven't looked hard at this case, but I doubt that there are any coherence assertions we can make for closed-categories-without-⊗ that are not implied by the solution to the coherence problem for $(\mathcal{B},\rho,\mathcal{D},\sigma)$ without (5.7). Moreover, if K is the club for this $(\mathcal{B},\rho,\mathcal{D},\sigma)$, I strongly suspect that K itself satisfies (5.7).

The theory of cartesian closed categories with finite limits and colimits can be expressed, after the appropriate extension from $\underline{\underline{P}}$ to $\underline{\underline{S}}^{op}$, or rather to $\underline{\underline{Cat}}^{op}$, in terms of a suitable $(\mathcal{B},\rho,\mathcal{D},\sigma)$. It would be interesting to know whether the theory of elementary topoi, <u>in those aspects related to coherence</u>, could also be so expressed; and whether the corresponding club K would then itself be an elementary topos. It is clear that K would be enormously complicated.

6. Free models - relation to Lambek and Szabo

6.1 We refer primarily to Lambek [9] on residuated categories and [10] on biclosed categories, and to Szabo ([17] and Ch.I of [15]) on closed categories; these all correspond to graphs in $\underline{\underline{P}}_0^*$; I have yet to examine in these terms Szabo's work on cartesian closed categories, which involves graphs in $\underline{\underline{S}}_o^{op*}$.

They start with a theory $(\mathcal{B},\rho,\mathcal{D},\sigma)$ as in §3, with ρ in fact vacuous in their cases, and construct directly the free model $K\circ A$ on A, but without mention of clubs, and indeed without mention of graphs. When it comes to a coherence result, therefore, they have to express it in other terms. Dealing with $K\circ A$ rather than K, they have variables from A in their morphisms, as in c: A⊗B → B⊗A instead of c: ⊗ → ⊗, and this to some extent serves to replace the concept of graph. But then to distinguish c, 1: A⊗A → A⊗A, where the variables no longer indicate the graph, they have to speak of these as admitting different "generalizations", namely c: A⊗B → B⊗A and 1: A⊗B → A⊗B; a coherence result we would state as "u = v if Γu = Γv" becomes for them roughly "u = v if u and v are equi-general".

<u>6.2</u> They construct the set $R = S\circ \underline{\underline{Ob}}\, A$ of objects of $K\circ A$ inductively, essentially as we constructed T(=S because ρ is vacuous) by (3.2) and (3.3), but with the objects of A as starting-point in place of our $\underline{\underline{1}}_\lambda$.

They then set up a <u>formal system</u> P whose formulae are T → S where T,S ∈ R. They give <u>axioms</u> and <u>rules of inference</u>; the axioms are in effect the maps f: A → B of A, the instances $d(T_1, \ldots, T_n)$ of the d ∈ $\mathcal{D}$ with $T_i \in R$, and 1: T → T for T ∈ R; the rules of inference are in effect, for B ∈ $\mathcal{B}$,

$$\frac{T_1 \to S_1 \qquad T_2 \to S_2 \qquad \ldots \qquad T_n \to S_n}{B(T_1, \ldots, T_n) \to B(S_1, \ldots, S_n)}$$

(appropriately altered when B is of mixed variance), and the <u>cut-rule</u>

(6.1) $$\frac{T \to R \qquad R \to S}{T \to S}.$$

(I say "in effect" because they may give an adjunction directly by

$$\frac{A\otimes B \to C}{A \to [B,C]} \quad \text{and} \quad \frac{A \to [B,C]}{A\otimes B \to C},$$

rather than, as we would, by its unit and counit as elements of $\mathcal{D}$; but it comes to the same thing.) They now consider for T, S $\in \mathcal{R}$ the set $\mathcal{P}(T,S)$ of all proofs of $T \to S$, written as trees; they describe an equivalence relation $\delta(T,S)$ on $\mathcal{P}(T,S)$, and form the quotient $\mathcal{P}(T,S)/\ \delta(T,S) = \mathcal{P}/\delta(T,S)$. Thus $\mathcal{P}/\delta$ is a precategory with $\mathcal{R}$ as objects. They define composition in $\mathcal{P}/\delta$ by using the cut, and show that it is actually a category, and indeed is the free model $K\circ A$ on A. The relation δ is chosen to achieve just this; it has to assert that the B $\in \mathcal{B}$ are functorial, that the d $\in \mathcal{D}$ are natural, and that the "imposed" relations σ are verified; it has to assert that composition via the cut is associative with identities 1_T; it has to assert that the composite in $\mathcal{P}/\delta$ of f,g $\in A$ really is their composite in A, and that 1_T is 1_A if $T \in A$; and finally that δ is an equivalence relation compatible with cut. It is not hard to see directly that this yields the same category as our $K\circ A$, or even as our $K[A]$ of §4.3 in the non-clubbable case; the main difference is that they allow the cut already in forming $\mathcal{P}$ - which otherwise would be in the case $A = I$ our Exp Inst $\mathcal{D}$ - so that their equivalence relation δ has to be more complicated than the Funct + Nat + Imp that we impose on Cat Exp Inst $\mathcal{D}$.

So long as the cut is one of the rules of inference, there is no obvious effective way of enumerating the proofs of T $\to$ S, for in (6.1) the term R may be arbitrarily more "complicated" than T or S. They overcome this by extending the cut-elimination technique in proof-theory due to Gentzen [4], a trick that Kelly-Mac Lane gratefully borrowed - without, however, the accompanying proof-theoretic language.

What Lambek and Szabo do, then, is to set up, still with T $\to$ S as formulae, a new formal system $\mathcal{Q}$, whose axioms are theorems of P, and whose rules of inference are derived rules in P. The rules of inference in $\mathcal{Q}$, however, are such that the terms in the hypotheses are parts of the terms in the thesis - in particular the cut is not a rule of inference in $\mathcal{Q}$ - and in favourable cases this allows an effective enumeration of the set $\mathcal{Q}(T,S)$ of proofs of $T \to S$ in $\mathcal{Q}$.

Assigning to each axiom of $\mathcal{Q}$ a particular proof of it in P, and to each rule of inference of $\mathcal{Q}$ a particular derivation of it in P, they get a function $\mathcal{Q}(T,S) \to P(T,S)$ called <u>expansion</u>; and they prove that the composite

(6.2) $\qquad \mathcal{Q}(T,S) \to P(T,S) \to P(T,S)/\delta(T,S)$

is surjective. Writing $\gamma(T,S)$ for the equivalence relation which is the kernel of (6.2), they now have $\mathcal{Q}(T,S)/\gamma(T,S) \cong P(T,S)/\delta(T,S)$; in other words the desired free model $K\circ A = P/\delta$ is also given by $\mathcal{Q}/\gamma$. An effective determination of $(K\circ A)(T,S)$ now only awaits an effective determination of $\gamma(T,S)$.

Observing that everything above is functorial in A, so that we can write $\mathcal{Q}(A)$ etc., Lambek and Szabo define on $\mathcal{Q}(A)(T,S)$ a new equivalence relation $\varepsilon(T,S)$ called <u>equi-generality</u>. Except in [17], the definition is essentially this: $f,g \in \mathcal{Q}(A)(T,S)$ are equi-general if, for all functors α: $A' \to A$, and for all T', $S' \in \mathcal{Q}(A')$ with $\mathcal{Q}(\alpha)\ T' = T$ and $\mathcal{Q}(\alpha)\ S' = S$, there is an $f' \in \mathcal{Q}(A')(T',S')$ with $\mathcal{Q}(\alpha)\ f' = f$ if and only if there is a $g' \in \mathcal{Q}(A')(T',S')$ with $\mathcal{Q}(\alpha)\ g' = g$. Because of the nature of the deductive system $\mathcal{Q}$, it is not necessary to take all A' here, but only those which are finite coproducts $n\underline{2}$ of the "arrow" category $\underline{2}$ (two objects 0, 1 and one non-identity map $0 \to 1$). In fact given T,S we need only try a finite number of n's, so that $\varepsilon(T,S)$ is a finitely-decidable relation.

It is easy to see that $f \equiv g(\gamma)$ implies $f \equiv g(\varepsilon)$. The converse is shown by Lambek and Szabo (see e.g. Lemma on p.117 of [10]) to be equivalent to the following, which we state for simplicity in our own language, using the fact that $\mathcal{Q}/\gamma = K \circ A$: <u>there is at most one map</u> $P[A_1, \ldots, A_p] \to Q[A_{p+1}, \ldots, A_{p+q}]$ <u>in</u> $K \circ n\underline{2}$, <u>where</u> $P, Q \in K$ <u>and</u> $A_i \in n\underline{2}$, <u>if</u> $A_1, \ldots, A_{p+q}$ <u>are all different</u>. When this is the case, $\varepsilon = \gamma$ and $K \circ A$ is calculated effectively as $\mathcal{Q}/\varepsilon$.

That $\gamma = \varepsilon$ is asserted in [9] (Proposition 5) and in [10] (Coherence theorem for BMM, p.116); also in [15], but in this case later corrected in [17]. In view of its equivalent form above, it in fact says exactly that $\Gamma: K \to \underline{P}^*_o$ is faithful; for if f, g: P → Q with $\Gamma f = \Gamma g$, and if $h_1, \ldots, h_n$ are the non-identity maps of $n\underline{2}$, we get by an appropriate choice of distinct $A_i \in n\underline{2}$ the two maps $f[h_1, \ldots, h_n]$ and $g[h_1, \ldots, h_n]$ from $P[A_1, \ldots, A_p]$ to $Q[A_{p+1}, \ldots, A_{p+q}]$. Thus in [10] and [15] this result is false, since Γ is certainly not faithful for the theories of biclosed or of closed categories. In [17] Szabo has **modified** [15] by replacing ε by a finer relation ε', which in that context he still calls "equi--generality", **and by asserting that** $\gamma = \varepsilon'$.

Another result given in each of the above references, under various names, is that the canonical functor $A \to K \circ A$ is full and faithful. In our language, this functor sends A to $\underline{1}[A]$ and f to $\underline{1}[f]$. It is certainly faithful since $\underline{1}[f] \neq \underline{1}[g]$ for $f \neq g$. It is full precisely when there is no non-identity map $h: \underline{1} \to \underline{1}$ in K. Since Γh is necessarily 1, this is so when Γ, restricted to the endomorphisms of $\underline{1}$ in K, is faithful; it is certainly so for closed categories by the results of [7], since $\underline{1}$ is a proper shape.

<u>7. Non-strict morphisms of K-categories</u>

<u>7.1</u> We just amplify a little here the indications given in §1.4; we hope at another time to study some of the problems suggested by

the concept.

Write **Fun** for the 2-category whose objects are functors $\phi: A \to A'$, whose morphisms $\phi \to \psi$ are 2-cells in **Cat** of the form

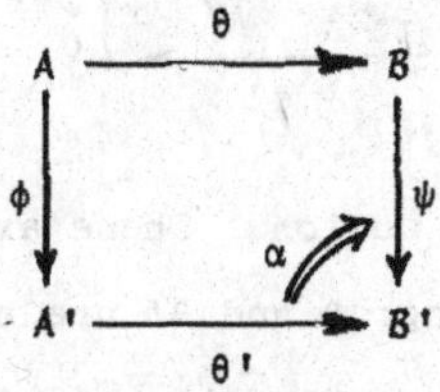

with the obvious composition, and whose 2-cells $(\theta,\theta',\alpha) \Rightarrow (\chi,\chi',\beta)$ are pairs of natural transformations $\gamma: \theta \Rightarrow \chi$ and $\gamma': \theta' \Rightarrow \chi'$ rendering commutative the following cylindrical diagram of 2-cells:

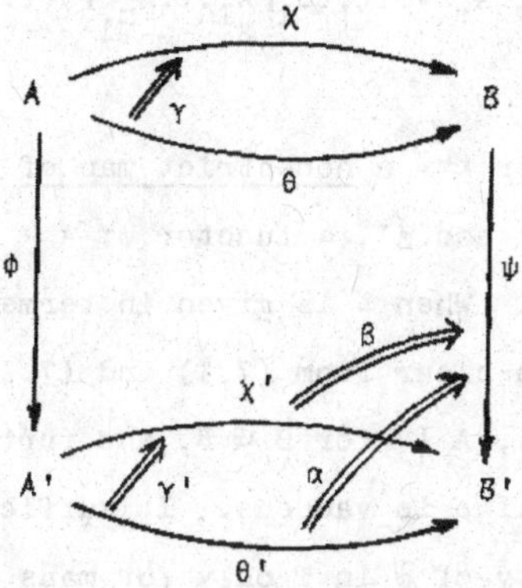

There is an obvious 2-functor $¤$: **Cat**/**P** × **Fun** → **Fun** sending the object (K,ϕ) to $K\circ\phi: K\circ A \to K\circ A'$, sending the morphism $(\rho,(\theta,\theta',\alpha))$ to $(\rho\circ\theta, \rho\circ\theta', \rho\circ\alpha)$, and sending the 2-cell $(\varepsilon,(\gamma,\gamma'))$ to $(\varepsilon\circ\gamma, \varepsilon\circ\gamma')$. This 2-functor satisfies the axioms for an <u>action</u> of **Cat**/**P** on **Fun**, so that we have $L¤(K¤\phi) = (L\circ K)¤\phi$, etc.

Now let K be a club in **Cat**/**P**. Then $K¤-$ is a monad (indeed a 2-monad) on **Fun**; an algebra for this monad is a functor $\phi: A \to A'$ together with an action $K¤\phi \to \phi$; that is, a 2-cell

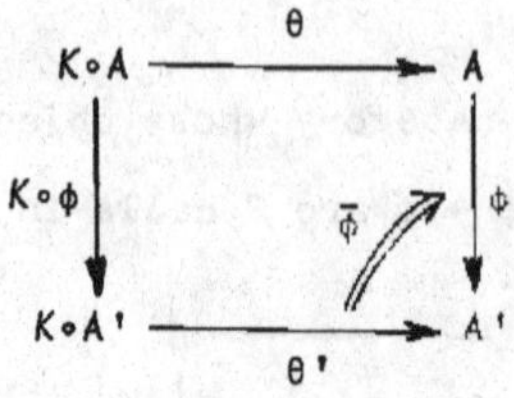

satisfying the axioms for an action. These axioms are easily seen to be equivalent to the following: θ and θ' are actions making A and A' into K-categories, while the components $\bar{\phi}(T[A_1, \ldots, A_n])$ satisfy (besides naturality in T and the A_i)

(7.1) $\bar{\phi}(T(S_1 \ldots S_n)[A_1 \ldots A_m])$

$= \bar{\phi}(T[S_1(A_1 \ldots A_{m_1}), \ldots, S_n(\ldots A_m)).T(\bar{\phi}(S_1[A_1 \ldots A_{m_1}]), \ldots, \bar{\phi}(S_n[\ldots A_m]))$

(7.2) $\bar{\phi}(\underline{1}[A]) = 1.$

We call an algebra for K¤- a <u>non-strict map of K-categories</u>; it consists of K-categories A and A', a functor $\phi\colon A \to A'$ and a natural $\bar{\phi}$ satisfying (7.1) and (7.2). When K is given in terms of generators and relations $(\mathcal{B}, \rho, \mathcal{D}, \sigma)$ it is clear from (7.1) and (7.2) that $\bar{\phi}$ is determined by the $\bar{\phi}(B[A_1, \ldots, A_n])$ for $B \in \mathcal{B}$, and further that these can be chosen independently if ρ is vacuous. It suffices moreover by (7.1) to assert the naturality of $\bar{\phi}$ in T only for maps $T \to T'$ in **Inst** $\mathcal{D}$. Thus we get, for instance, when K is the club for symmetric monoidal categories, the usual concept of a symmetric monoidal functor; $\bar{\phi}(\otimes[A,B])$ is a natural transformation $\tilde{\phi}\colon \phi A \otimes' \phi B \to \phi(A \otimes B)$ and $\bar{\phi}(I[\ ])$ is $\phi^{\circ}\colon I' \to \phi I$; asserting the naturality of $\bar{\phi}$ for instances of a,r,c gives the usual three axioms.

REFERENCES

[1] S.Eilenberg and G.M. Kelly, A generalization of the functorial calculus, J. Algebra 3(1966), 366-375.

[2] S.Eilenberg and G.M. Kelly, Closed Categories, in: Proc. Conf. on Categorical Algebra, La Jolla, 1965 (Springer-Verlag, 1966), 421-562.

[3] D.B.A. Epstein, Functors between tensored categories, Invent. Math. 1(1966), 221-228.

[4] G.Gentzen, Untersuchungen über das logische Schliessen I, II, Math.Z. 39(1934-1935), 176-210 and 405-431.

[5] G.M. Kelly, Many-variable functorial calculus I. (in this volume).

[6] G.M. Kelly, A cut-elimination theorem. (in this volume).

[7] G.M. Kelly and S. Mac Lane, Coherence in closed categories, J. Pure and Applied Algebra 1(1971), 97-140.

[8] G.M. Kelly and S. Mac Lane, Closed coherence for a natural transformation. (in this volume).

[9] J. Lambek, Deductive systems and categories I. Syntactic calculus and residuated categories, Math. Systems Theory 2(1968), 287-318.

[10] J. Lambek, Deductive systems and categories II. Standard constructions and closed categories, Lecture Notes in Mathematics 86(1969), 76-122.

[11] F.W. Lawvere, Ordinal sums and equational doctrines, Lecture Notes in Mathematics 80(1969), 141-155.

[12] G. Lewis, Coherence for a closed functor. (in this volume).

[13] J.L. Mac Donald, Coherence of adjoints, associativities, and identities, Arch. Math. 19(1968), 398-401.

[14] S. Mac Lane, Natural associativity and commutativity, Rice University Studies 49(1963), 28-46.

[15] M.E. Szabo, Proof-theoretical investigations in categorical algebra (Ph. D. Thesis, McGill Univ., 1970).

[16] M.E. Szabo, A categorical equivalence of proofs (to appear).

[17] M.E. Szabo, The logic of closed categories (to appear).

COHERENCE FOR A CLOSED FUNCTOR

Geoffrey Lewis

The University of New South Wales, Kensington 2033, Australia.

Received May 22, 1972

§0 Introduction

In the first published work on coherence, [6], Mac Lane showed coherence in tensored categories with and without identities, and with and without symmetry. In [2], Epstein showed coherence for a monoidal functor between two symmetric monoidal categories without identities. Kelly and Mac Lane in [5] examined coherence in a closed category.

We consider the structure consisting of two closed categories and a closed functor between them. The object of this paper is to prove partial coherence results as was done in [5] for a closed category.

As there, the purely covariant case is treated first: two symmetric monoidal categories and a monoidal functor. Then a cut-elimination theorem is proved, which leads to an inductive proof of partial coherence results. At the same time the cut-elimination proof will also yield a proof that the theory is clubbable as in [4]. The result will then be a partial determination of the club.

However the situation is worse than in [5] because even in the purely monoidal situation it is no longer the case that "all diagrams commute"; for example, when Φ is the forgetful functor from abelian groups to sets, the following diagram:

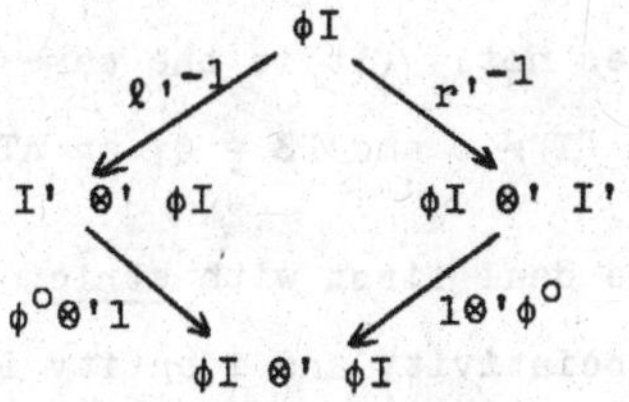

fails to commute. In this purely monoidal case we completely describe the club E. Like all the clubs we consider, E is a category over Λ, the discrete category with two objects A and B. The part of E with codomain type A is equivalent to $\underline{\underline{P}}$. The part of E with codomain type B is equivalent to the category H^*. An object of H^* is (n, p, u, x, α); where n, p, u are integers ≥ 0, $x: n \to p$ is an increasing map, and α is a (p,u) - shuffle. There are morphisms from (n, p, u, x, α) to $(n', p', u', x', \alpha')$ only when $n = n'$, $u = u'$. Such a morphism f is (ξ, h, θ); where $\xi: n \to n$ and $\theta: u \to u$ are permutations, $h: p \to p'$ is a function, and $x'\xi = hx$. The permutations ξ and θ together constitute the graph Γf in the sense of Kelly [3]; h is a new characteristic of f, that we shall call Δf.

We thus have $\Gamma: H^* \to \underline{\underline{Q}} = \underline{\underline{P}} \circ \Lambda \times \Lambda$. We also have a new functor $\Delta: H^* \to \underline{\underline{\Delta}}$, where $\underline{\underline{\Delta}}$ is the skeletal category of finite sets. While Γ is not faithful, Γ and Δ are jointly faithful, i.e. $f = g$ iff $\Gamma f = \Gamma g$ and $\Delta f = \Delta g$.

When we come to the club N for the closed functor, we still have a graph function Γ, but we depart from §4. 2-3 of [4] by letting it take its value in a modification G of $\underline{\underline{P}}^*_o$. We define a category D which bears the same relation to $\underline{\underline{\Delta}}$ as G does to $\underline{\underline{Q}}$, and we extend $\Delta: H^* \to \underline{\underline{\Delta}}$ to $\Delta: N \to D$. In the case of N, it is not true that Γ and Δ are jointly faithful: we do not attempt a complete description of the club N. We content ourselves with proving that Γ and Δ <u>are</u> jointly faithful when restricted to the full subcategory of N defined by its <u>proper</u> objects. An object is called proper if in its formation no

$[T,S]$ is involved, where either T is non-constant and S is constant, or T contains ϕ and S does not. (It is the same thing to say that $[T,S]$ is never used with $\Gamma T \neq 0$ and $\Gamma S = 0$, or $\Delta T \neq 0$ and $\Delta S = 0$).

For simplicity we deal first with strict symmetric monoidal categories (i.e. all associativity and identity isomorphisms are identity morphisms), and to facilitate this we first consider strict non-symmetric monoidal categories.

§1 Determination of the club in the strict non-symmetric monoidal case

We seek the club F, in the sense of [4] for the system given by:

Two strict non-symmetric monoidal categories $V = \{A, \otimes, I, a, \ell, r\}$ and $V' = \{A', \otimes', I', a', \ell', r'\}$, and a monoidal functor $\Phi = \{\phi, \tilde{\phi}, \phi^o\}: V \to V'$, as in [1].

We write F_A and F_B for the parts of F of codomain types 1 and 2, or, as we shall say, of types A and B. Type A objects can be identified with the non-negative integers. Type B objects are of the form:

$$C_1 \otimes' C_2 \otimes' \dots \otimes' C_m$$

where C_i is either $\underline{1}'$ or ϕn_i. If $m = 0$, the object is written I'.

Type A morphisms are $1: n \to n$. Type B morphisms are generated by expanded instances of

$$\tilde{\phi}: \phi n \otimes' \phi m \to \phi(n + m), \text{ and}$$

$$\phi^o: I' \to \phi 0.$$

[NB: In expanded instances of ϕ^o, I' is understood, e.g. $1 \otimes' \phi^o \otimes' 1: \underline{1}' \otimes' \phi n \to \underline{1}' \otimes' \phi 0 \otimes' \phi n$]

The generators satisfy functorial relations:

$$\begin{array}{ccc}
D_1 \otimes' D_2 \otimes' D_3 \otimes' D_4 \otimes' D_5 & \xrightarrow{1 \otimes' f \otimes' 1} & D_1 \otimes' E_2 \otimes' D_3 \otimes' D_4 \otimes' D_5 \\
\downarrow{\scriptstyle 1 \otimes' g \otimes' 1} & & \downarrow{\scriptstyle 1 \otimes' g \otimes' 1} \\
D_1 \otimes' D_2 \otimes' D_3 \otimes' E_4 \otimes' D_5 & \xrightarrow[1 \otimes' f \otimes' 1]{} & D_1 \otimes' E_2 \otimes' D_3 \otimes' E_4 \otimes' D_5 ;
\end{array}$$

naturality relations (vacuous in this club); and the given relations:

$$\begin{array}{ccccc}
 & & D \otimes' \phi n \otimes' \phi m \otimes' \phi p \otimes' & & \\
 & {\scriptstyle 1\otimes'\tilde{\phi}\otimes'1} \swarrow & & \searrow {\scriptstyle 1\otimes'\tilde{\phi}\otimes'1} & \\
D \otimes' \phi(n+m) \otimes' \phi p \otimes' E & & & & D \otimes' \phi n \otimes' \phi(m+p) \otimes' E \\
 & {\scriptstyle 1\otimes'\tilde{\phi}\otimes'1} \searrow & & \swarrow {\scriptstyle 1\otimes'\tilde{\phi}\otimes'1} & \\
 & & D \otimes' \phi(n+m+p) \otimes' E & &
\end{array}$$

$$\begin{array}{ccccc}
 & & D \otimes' \phi n \otimes' E & & \\
 & {\scriptstyle 1\otimes'\phi^{0}\otimes'1} \swarrow & & \searrow {\scriptstyle 1\otimes'\phi^{0}\otimes'1} & \\
D \otimes' \phi 0 \otimes' \phi n \otimes' E & & \downarrow {\scriptstyle 1} & & D \otimes' \phi n \otimes' \phi 0 \otimes' E \\
 & {\scriptstyle 1\otimes'\tilde{\phi}\otimes'1} \searrow & & \swarrow {\scriptstyle 1\otimes'\tilde{\phi}\otimes'1} & \\
 & & D \otimes' \phi n \otimes' E & &
\end{array}$$

We shall describe a category H. The objects of H are either n or $n \xrightarrow{f} p \xrightarrow{a} u$, where n,p,u are non-negative integers, and $f:[n] \to [p]$ and $a: [p] \to [u]$ are increasing functions. $[0]$ is the empty set, and $[n]$ is the set $\{1, 2, \ldots, n\}$ for $n > 0$. Henceforth we shall omit the square brackets in this context. The morphisms of H are $1: n \to n$ and

$$\begin{array}{ccccc}
n & \xrightarrow{f} & p & \xrightarrow{a} & u \\
\downarrow{\scriptstyle 1} & & \downarrow{\scriptstyle h} & & \downarrow{\scriptstyle 1} \\
n & \xrightarrow[g]{} & q & \xrightarrow[b]{} & u
\end{array} \qquad \text{(sometimes written } (1,h,1))$$

where h is increasing, such that the diagram commutes. Composition is defined in the obvious way so that

$$(1, k, 1).\ (1, h, 1) = (1, kh, 1)$$

<u>Theorem 1.1</u>: <u>$\mathcal{F}$ is isomorphic to $\mathcal{H}$</u>

<u>Proof</u>: Let $\mathcal{B}$ be the category with the type B objects of $\mathcal{F}$ as objects, and such that $\mathcal{B}$ is the free category on generators expanded instances of $\tilde{\phi}$ and ϕ^{o}. Denote the relations on these generators in $\mathcal{F}$ by ρ; the congruence thus generated by $[\rho]$. Then $\mathcal{F}_B = \mathcal{B}/[\rho]$. We want to show that $\mathcal{B}/[\rho] \simeq \mathcal{H}_B$, where $\mathcal{H}_B$ is the subcategory of $\mathcal{H}$ without the objects n, and the morphisms $1: n \to n$. We shall exhibit $\mathcal{H}_B$ as $\mathcal{B}'/[\rho']$ where $\mathcal{B}' \simeq \mathcal{B}$ and $\rho' \simeq \rho$.

Consider the object D of $\mathcal{B}$:

$$\phi n_1 \otimes' \dots \otimes' \phi n_{r_1} \otimes' \underline{1}' \otimes' \phi n_{r_1+1} \otimes' \dots \otimes' \phi n_{r_2} \otimes \underline{1}' \otimes \dots$$

$$\dots \otimes \underline{1}' \otimes' \phi n_{r_{u-1}+1} \otimes' \dots \otimes' \phi n_{r_u}$$

Let $p = r_u$ and $n = \sum_{i=1}^{p} n_i$. Define $f: n \to p$ by $f(i) = j$ if $n_1 + \dots + n_{j-1} < i \leq n_1 + \dots + n_j$. Define $a: p \to u$ by $a(i) = j$ if $r_{j-1} < i \leq r_j$. Then there is a bijection between the objects of $\mathcal{B}$ and $\mathcal{H}_B$ given by the correspondence between D and $n \xrightarrow{f} p \xrightarrow{a} u$.

Let $\sigma^i: p \to p-1$ be the increasing surjection which takes the value i twice. Let $\delta^i: p \to p+1$ be the increasing injection which fails to take the value i. Denote by $\sigma^i_{f,b}$ and $\delta^i_{f,b}$, respectively, the following morphisms in $\mathcal{H}_B$:

$$\begin{array}{ccccc} n & \xrightarrow{f} & p & \xrightarrow{b\sigma^i} & u \\ {\scriptstyle 1}\downarrow & & \downarrow{\scriptstyle \sigma^i} & & \downarrow{\scriptstyle 1} \\ n & \xrightarrow[\sigma^i f]{} & p-1 & \xrightarrow[b]{} & u \end{array} \quad \text{and} \quad \begin{array}{ccccc} n & \xrightarrow{f} & p & \xrightarrow{b\delta^i} & u \\ {\scriptstyle 1}\downarrow & & \downarrow{\scriptstyle \delta^i} & & \downarrow{\scriptstyle 1} \\ n & \xrightarrow[\delta^i f]{} & p+1 & \xrightarrow[b]{} & u \end{array} .$$

Let $\mathcal{B}'$ have the same objects as $\mathcal{H}_B$, and be the free category on generators $\sigma^i_{f,b}$ and $\delta^i_{f,b}$. Identify $D \otimes' \phi n_i \otimes' \phi n_{i+1} \otimes' E$ with $\sigma^i_{f,b}$

$$\begin{array}{c} D \otimes' \phi n_i \otimes' \phi n_{i+1} \otimes' E \\ \downarrow {\scriptstyle 1 \otimes' \tilde{\phi} \otimes' 1} \\ D \otimes' \phi(n_i + n_{i+1}) \otimes' E \end{array}$$

and
$$\begin{array}{c} D \otimes' E \\ \big\downarrow {\scriptstyle 1 \otimes' \phi^{0} \otimes' 1} \\ D \otimes' \phi 0 \otimes' E \end{array} \qquad \text{with } \delta^{i}_{f,b}.$$

This is a bijection between the generators of $\mathcal{B}$ and of $\mathcal{B}'$. The isomorphism between $\mathcal{B}$ and $\mathcal{B}'$ sends ρ to the isomorphic relations ρ' in $\mathcal{B}'$. The problem reduces to showing $\mathcal{B}'/[\rho'] \approx \mathcal{H}_B$. The point of the proof is that the relations ρ' have the same form as the relations τ below.

Define a functor $K\colon \mathcal{B}' \to \mathcal{H}_B$ by the bijections just described on the objects and generators. Let $\underline{\Delta}$ be the category with objects $n \geq 0$, and morphisms increasing maps. Define a functor $F\colon \mathcal{H}_B \to \underline{\Delta}$ by

$$F(n \xrightarrow{f} p \xrightarrow{a} u) = p, \qquad \text{and}$$

$$F\left(\begin{array}{ccccc} n & \xrightarrow{f} & p & \xrightarrow{a} & u \\ {\scriptstyle 1}\big\downarrow & & \big\downarrow{\scriptstyle h} & & \big\downarrow{\scriptstyle 1} \\ n & \xrightarrow[g]{} & q & \xrightarrow[b]{} & u \end{array}\right) = h$$

It is well known that $\underline{\Delta}$ has generators $\sigma^{i}_{p}\colon p \to p-1$ and $\delta^{i}_{p}\colon p \to p+1$ and relations τ:

$$\begin{array}{lll} \delta^{i}_{p-1}\sigma^{i}_{p} = & \sigma^{i}_{p+1}\,\delta^{j+1}_{p} & j > i \\ & \sigma^{i+1}_{p+1}\,\delta^{j}_{p} & j < i \\ & 1 & j = i \\ \sigma^{j}_{p-1}\,\sigma^{i}_{p} = & \sigma^{i}_{p-1}\,\sigma^{j+1}_{p} & j \geq i \\ \delta^{j}_{p+1}\,\delta^{i}_{p} = & \delta^{i}_{p+1}\,\delta^{j-1}_{p} & j \geq i \end{array}$$

Let C be the category with objects $n \geq 0$, and morphisms freely generated by σ^{i}_{p} and δ^{i}_{p}. Let $L\colon C \to \underline{\Delta}$ be the functor which is the identity on objects and generators. There is a functor $G\colon \mathcal{B}' \to C$ which is the identity on objects and such that $G(\sigma^{i}_{f,b}) = \sigma^{i}_{p}$ and $G(\delta^{i}_{f,b}) = \delta^{i}_{p}$. We have $FK = LG$.

We know that ker L = $[\tau]$. We are required to prove that K is onto, and that ker K = $[\rho']$.

Choose a morphism, α, of $\mathcal{H}_B$:

$$\begin{array}{ccccc} n & \xrightarrow{f} & p & \xrightarrow{bh} & u \\ {\scriptstyle 1}\downarrow & & \downarrow{\scriptstyle h} & & \downarrow{\scriptstyle 1} \\ n & \xrightarrow[hf]{} & q & \xrightarrow[b]{} & u. \end{array}$$

Since $h \in \underline{\Delta}$, h may be written as

$$p = p_o \xrightarrow{h_1} p_1 \xrightarrow{h_2} p_2 \ldots \xrightarrow{h_m} p_m = q$$

for generators h_i of $\underline{\Delta}$. But α is the composite:

$$(*) \qquad \begin{array}{ccccc} n & \xrightarrow{f} & p & \xrightarrow{bh_m \ldots h_1 = bh} & u \\ {\scriptstyle 1}\downarrow & & \downarrow{\scriptstyle h_1} & & \downarrow{\scriptstyle 1} \\ n & \xrightarrow{h_1 f} & p_1 & \xrightarrow{bh_m \ldots h_2} & u \\ {\scriptstyle 1}\downarrow & & \downarrow{\scriptstyle h_2} & & \downarrow{\scriptstyle 1} \\ n & \xrightarrow{h_2 h_1 f} & p_2 & \xrightarrow{bh_m \ldots h_3} & u \\ \vdots & & \vdots & & \vdots \\ n & \xrightarrow{h_{m-1} \ldots h_1 f} & p_{m-1} & \xrightarrow{bh_m} & u \\ {\scriptstyle 1}\downarrow & & \downarrow{\scriptstyle h_m} & & \downarrow{\scriptstyle 1} \\ n & \xrightarrow{h_m \ldots h_1 f} & p_m & \xrightarrow{b} & u. \end{array}$$

So K is clearly onto.

Suppose $f \simeq g\ [\rho']$. Then $Gf \simeq Gg\ [\tau]$. So LGf = LGg, thus FKf = FKg. But F is faithful, so Kf = Kg. Thus ker K $\supset [\rho']$.

If Kf = Kg, then FKf = FKg, so LGf = LGg, and $Gf \simeq Gg\ [\tau]$. But by the decomposition in (*), $f \simeq g\ [\rho']$. Thus ker K = $[\rho']$. QED.

§2 Determination of the club in the strict symmetric monoidal case

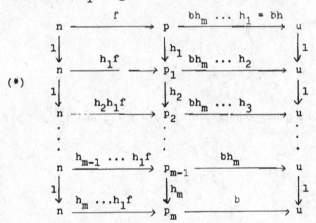

We seek to describe the club F^* for a monoidal functor between strict symmetric monoidal categories. We first describe F', which is the full subcategory of F^* determined by the objects not involving $\underline{1}'$.

Type A objects of F' are non-negative integers n, and type B objects are $\phi n_1 \otimes' \phi n_2 \otimes' \dots \otimes' \phi n_p$. (For convenience we shall omit the prime and write this as $\phi n_1 \otimes \phi n_2 \otimes \dots \otimes \phi n_p$). Type A morphisms are $\xi: n \to n$ where $\xi \in S(n)$, the permutation group on n elements. Type B morphisms are generated by:

(1) Type B morphisms of F whose domain and range do not involve $\underline{1}'$;

(2) $\phi n_{\xi 1} \otimes \dots \otimes \phi n_{\xi p} \xrightarrow{\hat{\xi}} \phi n_1 \otimes \dots \otimes \phi n_p$, where $\xi \in S(p)$;

(3) $\phi\xi: \phi n \to \phi n$, where $\xi \in S(n)$;

The relations are:

(1) Functorial:

$$\begin{array}{ccc} A \otimes B \otimes C \otimes D \otimes E & \xrightarrow{1 \otimes f \otimes 1} & A \otimes B' \otimes C \otimes D \otimes E \\ {\scriptstyle 1 \otimes g \otimes 1}\downarrow & & \downarrow{\scriptstyle 1 \otimes g \otimes 1} \\ A \otimes B \otimes C \otimes D' \otimes E & \xrightarrow[1 \otimes f \otimes 1]{} & A \otimes B' \otimes C \otimes D' \otimes E \end{array}$$

(2) Natural:

(a)
$$\begin{array}{ccc} A \otimes \phi n \otimes \phi m \otimes B & \xrightarrow{1 \otimes \phi\xi \otimes \phi\eta \otimes 1} & A \otimes \phi n \otimes \phi m \otimes B \\ {\scriptstyle 1 \otimes \tilde{\phi} \otimes 1}\downarrow & & \downarrow{\scriptstyle 1 \otimes \tilde{\phi} \otimes 1} \\ A \otimes \phi(n+m) \otimes B & \xrightarrow[1 \otimes \phi(2(\xi,\eta)) \otimes 1]{} & A \otimes \phi(n+m) \otimes B \end{array}$$

(b)
$$\begin{array}{ccc} A_{\xi 1} \otimes \dots \otimes A_{\xi p} & \xrightarrow{\hat{\xi}(1_{A_1},\dots,1_{A_p})} & A_1 \otimes \dots \otimes A_p \\ \downarrow{\scriptstyle f_{\xi 1} \otimes \dots \otimes f_{\xi p}} & & \downarrow{\scriptstyle f_1 \otimes \dots \otimes f_p} \\ B_{\xi 1} \otimes \dots \otimes B_{\xi p} & \xrightarrow[\hat{\xi}(1_{B_1},\dots,1_{B_p})]{} & B_1 \otimes \dots \otimes B_p \end{array}$$

(3) Given relations:

(a) If h, k are composable in $\mathcal{F}$, then the composite in $\mathcal{F}'$ is the composite in $\mathcal{F}$; similarly for identities.

(b) $\widehat{\xi\eta} = \hat{\xi}\hat{\eta}$

(c) $\phi(\xi\eta) = \phi\xi . \phi\eta$

(d)
$$\begin{array}{ccc}
A \otimes \phi n \otimes \phi m \otimes B & \xrightarrow{1 \otimes \hat{\tau} \otimes 1} & A \otimes \phi m \otimes \phi n \otimes B \\
{\scriptstyle 1 \otimes \tilde{\phi} \otimes 1}\downarrow & & \downarrow{\scriptstyle 1 \otimes \tilde{\phi} \otimes 1} \\
A \otimes \phi(n+m) \otimes B & \xrightarrow[1 \otimes \phi(\tau(n,m)) \otimes 1]{} & A \otimes \phi(n+m) \otimes B
\end{array}$$

where τ is the non-identity element of S(2).

$\mathcal{F}'_A$ is isomorphic to the category $\underline{\underline{P}}$. (Objects n, $\underline{\underline{P}}$ (n,m) = $\emptyset$ if $n \neq m$, $\underline{\underline{P}}$ (n,n) = S(n)). We shall show that $\mathcal{F}'_B$ is isomorphic to the category $\mathcal{H}'$.

The objects of $\mathcal{H}'$ are increasing maps $f: n \to p$. A morphism of $\mathcal{H}'$ is a commutative diagram

$$\begin{array}{ccc}
n & \xrightarrow{f} & p \\
{\scriptstyle \xi}\downarrow & & \downarrow{\scriptstyle h} \\
n & \xrightarrow[g]{} & q
\end{array}$$

where $\xi \in S(n)$, and h is any function. We sometimes write this as (ξ,h). The composite with (ξ', h') is $(\xi'\xi, h'h)$.

We shall now study the morphisms of $\mathcal{H}'$. Given an increasing $f: n \to p$, and a permutation $\eta: p \to p$ we write as η^f the permutation of n (in the notation of [3] §2):

$$\eta(1_{a_{\eta^{-1}1}}, \ldots, 1_{a_{\eta^{-1}p}}): p(a_1, \ldots, a_p) \to p(a_{\eta^{-1}1}, \ldots, a_{\eta^{-1}p})$$

where $a_i = f^{-1}i$. In (1) h may be written as $k\eta$ where $\eta \in S(p)$ and

$k \in \underline{\Delta}(p,q)$. Then (1) may be written as :

$$(2)\qquad \begin{array}{ccc} n & \xrightarrow{\quad f \quad} & p \\ {\scriptstyle \eta^f}\big\downarrow & & \big\downarrow{\scriptstyle \eta} \\ n & \xrightarrow{\eta f.(\eta^f)^{-1}} & p \\ {\scriptstyle 1}\big\downarrow & & \big\downarrow{\scriptstyle k} \\ n & \xrightarrow{k\eta f.(\eta^f)^{-1}} & q \\ {\scriptstyle \xi(\eta^f)^{-1}}\big\downarrow & & \big\downarrow{\scriptstyle 1} \\ n & \xrightarrow{k\eta f\xi^{-1} = g} & q \end{array}$$

This is a composite of three morphisms of $H!$ and is uniquely determined by η and k. We examine what happens when h may also be written as $k'\eta'$ for $\eta' \in S(p)$, $k' \in \underline{\Delta}(p,q)$.

Let $\eta^* = \eta'\eta^{-1}$. Thus $k = k'\eta^*$. Suppose $i < j$. If $ki < kj$, then $k'\eta^*i < k'\eta^*j$, so $\eta^*i < \eta^*j$ because k' is increasing. If $ki = kj$, then $k'\eta^*i = k'\eta^*j$, so η^*i may be either less than or greater than η^*j. It follows that $\eta^* = 1_q(\eta_1, \ldots, \eta_q): p \to p$ where $\eta_i \in S(k^{-1}(i))$. For any $i \in p$, $k'i = (k'(\eta^*)^{-1}(\eta^*))i = k(\eta^*i) = ki$. Thus $k = k'$.

We define $\otimes$ in H:

$(f_1: n_1 \to p_1) \otimes (f_2: n_2 \to p_2) = 2(f_1,f_2): 2(n_1,n_2) \to 2(p_1,p_2)$

$$\left(\begin{array}{ccc} n_1 & \xrightarrow{f_1} & p_1 \\ {\scriptstyle \xi_1}\big\downarrow & & \big\downarrow{\scriptstyle h_1} \\ n_1 & \xrightarrow[g_1]{} & q_1 \end{array}\right) \otimes \left(\begin{array}{ccc} n_2 & \xrightarrow{f_2} & p_2 \\ {\scriptstyle \xi_2}\big\downarrow & & \big\downarrow{\scriptstyle h_2} \\ n_2 & \xrightarrow[g_2]{} & q_2 \end{array}\right) = \begin{array}{ccc} 2(n_1,n_2) & \xrightarrow{2(f_1,f_2)} & 2(p_1,p_2) \\ \big\downarrow{\scriptstyle 2(\xi_1,\xi_2)} & & \big\downarrow{\scriptstyle 2(h_1,h_2)} \\ 2(n_1,n_2) & \xrightarrow[2(g_1,g_2)]{} & 2(q_1,q_2) \end{array}$$

$\otimes$ is strictly associative.

By the decomposition (2) we note that there are three types of generators of morphisms of H':

type (C):

(3)
$$\begin{array}{ccc} n & \xrightarrow{\quad f \quad} & p \\ {\scriptstyle \eta^f}\downarrow & & \downarrow{\scriptstyle \eta} \\ n & \xrightarrow[\eta f.(\eta^f)^{-1}]{} & p \end{array}$$

type (D):

(4)
$$\begin{array}{ccc} n & \xrightarrow{\quad f \quad} & p \\ {\scriptstyle 1}\downarrow & & \downarrow{\scriptstyle h} \\ n & \xrightarrow[hf]{\qquad} & q \end{array}$$

type (E):

(5)
$$\begin{array}{ccc} n & \xrightarrow{\quad f \quad} & p \\ {\scriptstyle \xi}\downarrow & & \downarrow{\scriptstyle 1} \\ n & \xrightarrow[f]{\qquad} & p \end{array}$$

where ξ may be written as

$\hat{1}(\xi_1, \ldots, \xi_p)\colon p(a_1, \ldots, a_p) \to p(a_1, \ldots, a_p)$

where $a_i = f^{-1}(i)$, and $\xi_i \in S(a_i)$.

Clearly $(\eta_1{}^g, \eta_1).(\eta^f, \eta) = (\eta_1\eta^f, \eta_1\eta)$ where $g = \eta f(\eta^f)^{-1}$; $(1, h_1).\ (1, h) = (1, h_1 h)$; and $(\xi', 1).(\xi, 1) = (\xi'\xi, 1)$.

We now prove:

<u>Lemma 2.1</u>: <u>If</u> α <u>is of type</u> (*), <u>and</u> β <u>is of type</u> (†), <u>then there exists</u> α' <u>of type</u> (*), <u>and</u> β' <u>of type</u> (†), <u>such that</u> $\alpha\beta = \beta'\alpha'$, <u>where</u>

(1) * <u>is</u> C, † <u>is</u> D

(2) * <u>is</u> C, † <u>is</u> E

(3) * <u>is</u> D, † <u>is</u> E

<u>Note</u> We extend the notations $1(\xi_1, \ldots, \xi_p)$ and $\eta(1, 1, \ldots, 1)$ of [3] §2 in an obvious way to the case where ξ, η are functions and no longer merely permutations.

<u>**Proof**</u>

(1) $\alpha\beta =$

$$\begin{CD}
n @>{f}>> p \\
@V{1}VV @VV{h}V \\
n @>{hf}>> q \\
@V{\eta^{hf}}VV @VV{\eta}V \\
n @>>{\eta hf(\eta^{hf})^{-1}}> q
\end{CD}$$

$$\begin{CD}
n = q(n_{\eta 1}, \ldots, n_{\eta q}) @>{f = 1(f_{\eta 1}, \ldots, f_{\eta q})}>> q(p_{\eta 1}, \ldots, p_{\eta q}) = p \\
@V{1}VV @VV{1(h_{\eta 1}, \ldots, h_{\eta q}) = h}V \\
q(n_{\eta 1}, \ldots, n_{\eta q}) @>{1(h_{\eta 1}f_{\eta 1}, \ldots, h_{\eta q}f_{\eta q})}>> q(1, \ldots, 1) \\
@V{\eta(1, \ldots, 1)}VV @VV{\eta(1, \ldots, 1)}V \\
q(n_1, \ldots, n_q) @>{1(h_1f_1, \ldots, h_qf_q)}>> q(1, \ldots, 1)
\end{CD}$$

$=$

$$\begin{CD}
q(n_{\eta 1}, \ldots, n_{\eta q}) @>{1(f_{\eta 1}, \ldots, f_{\eta q})}>> q(p_{\eta 1}, \ldots, p_{\eta q}) \\
@V{\eta(1, \ldots, 1)}VV @VV{\eta(1, \ldots, 1)}V \\
q(n_1, \ldots, n_q) @>{1(f_1, \ldots, f_q)}>> q(p_1, \ldots, p_q) \\
@V{1}VV @VV{1(h_1, \ldots, h_q)}V \\
q(n_1, \ldots, n_q) @>{1(h_1f_1, \ldots, h_qf_q)}>> q(1, \ldots, 1)
\end{CD}$$

$= \beta'\alpha'$

(2) $\alpha\beta =$

$$\begin{CD}
n @>{f}>> p \\
@V{\xi}VV @VV{1}V \\
n @>{f}>> p \\
@V{\eta^{f}}VV @VV{\eta}V \\
n @>>{\eta f.(\eta^{f})^{-1}}> p
\end{CD}$$

$$\begin{array}{ccc}
n = p(n_{\eta 1}, \ldots, n_{\eta p}) & \xrightarrow{f \,=\, 1(f_{\eta 1}, \ldots, f_{\eta p})} & p(1, \ldots, 1) \\
\Big\downarrow {\scriptstyle \xi \,=\, 1(\xi_{\eta 1}, \ldots, \xi_{\eta p})} & & \Big\downarrow {\scriptstyle 1} \\
p(n_{\eta 1}, \ldots, n_{\eta p}) & \xrightarrow{1(f_{\eta 1}, \ldots, f_{\eta p})} & p(1, \ldots, 1) \\
\Big\downarrow {\scriptstyle \eta(1, \ldots, 1)} & & \Big\downarrow {\scriptstyle \eta(1, \ldots, 1)} \\
p(n_1, \ldots, n_p) & \xrightarrow{1(f_1, \ldots, f_p)} & p(1, \ldots, 1)
\end{array}$$

$$= \quad \begin{array}{ccc}
p(n_{\eta 1}, \ldots, n_{\eta p}) & \xrightarrow{1(f_{\eta 1}, \ldots, f_{\eta p})} & p(1, \ldots, 1) \\
\Big\downarrow {\scriptstyle \eta(1, \ldots, 1)} & & \Big\downarrow {\scriptstyle \eta(1, \ldots, 1)} \\
p(n_1, \ldots, n_p) & \xrightarrow{1(f_1, \ldots, f_p)} & p(1, \ldots, 1) \\
\Big\downarrow {\scriptstyle 1(\eta_1, \ldots, \eta_p)} & & \Big\downarrow {\scriptstyle 1} \\
p(n_1, \ldots, n_p) & \xrightarrow{1(f_1, \ldots, f_p)} & p(1, \ldots, 1)
\end{array}$$

$= \beta'\alpha'$

(3) $\alpha\beta =$

$$\begin{array}{ccc}
n & \xrightarrow{f} & p \\
\Big\downarrow {\scriptstyle \xi} & & \Big\downarrow {\scriptstyle 1} \\
n & \xrightarrow{f} & p \\
\Big\downarrow {\scriptstyle 1} & & \Big\downarrow {\scriptstyle h} \\
n & \xrightarrow[hf]{} & q
\end{array}$$

$$
= \quad
\begin{array}{ccc}
p(n_1, \ldots, n_p) & \xrightarrow{\;p(f_1, \ldots, f_p)\;} & p(1, \ldots, 1) \\
\downarrow{\scriptstyle 1(\xi_1, \ldots, \xi_p)} & & \downarrow{\scriptstyle 1} \\
p(n_1, \ldots, n_p) & \xrightarrow{\;p(f_1, \ldots, f_p)\;} & p(1, \ldots, 1) \\
\Big\downarrow{\scriptstyle 1} & & \downarrow{\scriptstyle =} \\
 & & (q(p_1, \ldots, p_q))(1, \ldots, 1) \\
 & & \Big\downarrow{\scriptstyle h \,=\, h(1, \ldots, 1) \,=\, (1(h_1, \ldots, h_q))(1, \ldots, 1)} \\
(q(p_1, \ldots, p_q))(n_1, \ldots, n_p) & \xrightarrow{\;(1(h_1, \ldots, h_q))(f_1, \ldots, f_p)\;} & (q(1, \ldots, 1))(1, \ldots, 1)
\end{array}
$$

$$
= \quad
\begin{array}{ccc}
p(n_1, \ldots, n_p) & \xrightarrow{\;1(f_1, \ldots, f_p)\;} & p(1, \ldots, 1) \\
\Big\downarrow{\scriptstyle 1} & & \downarrow{\scriptstyle =} \\
 & & (q(p_1, \ldots, p_q))(1, \ldots, 1) \\
 & & \downarrow{\scriptstyle (1(h_1 .. , h_q))(1 .. , 1)} \\
(q(p_1, \ldots, p_q))(n_1, \ldots, n_p) & \xrightarrow{\;(1(h_1, \ldots, h_q))(f_1 \ldots, f_p)\;} & (q(1, \ldots, 1))(1, \ldots, 1) \\
\downarrow{\scriptstyle =} & & \\
q(p_1(n_{11}, \ldots, n_{1p_1}), \ldots, p_q(n_{q1}, \ldots, n_{qp_q})) & & \\
\downarrow{\scriptstyle 1(1(\eta_{11}, \ldots, \eta_{1p_1}), \ldots, (1(\eta_{q1}, \ldots, \eta_{qp_q}))} & & \Big\downarrow{\scriptstyle 1} \\
q(p_1(n_{11}, \ldots, n_{1p_1}), \ldots, p_q(n_{q1}, \ldots, n_{qp_q})) & & \\
\downarrow{\scriptstyle =} & & \\
(q(p_1, \ldots, p_q))(n_1, \ldots, n_p) & \xrightarrow{\;(1(h_1, \ldots, h_q))(f_1, \ldots, f_p)\;} & (q(1, \ldots, 1))(1, \ldots, 1)
\end{array}
$$

$= \beta'\alpha'$; where $n_{11}, \ldots, n_{1p_1}, \ldots, n_{q1}, \ldots, n_{qp_q}$ is a relabelling of $n_1, \ldots, n_p$, and similarly for η.

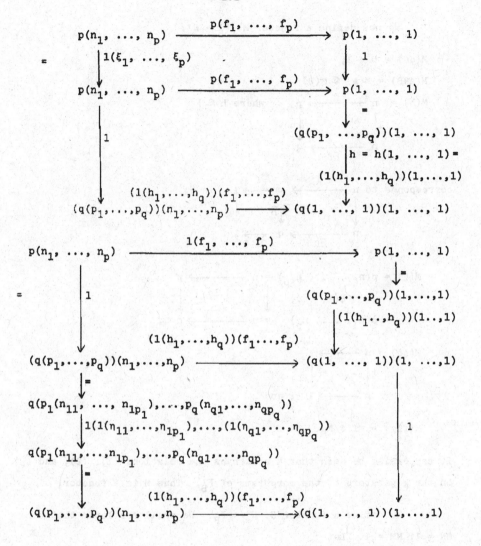

We now define a functor $M\colon F'_B \to H'$:

$M(\phi n) = n + 1;$

$M(A \otimes B) = M(A) \otimes M(B);$

$$M(h) = \begin{array}{ccc} n & \xrightarrow{\ f\ } & p \\ {\scriptstyle 1}\downarrow & & \downarrow{\scriptstyle \bar{h}} \\ n & \xrightarrow[\ \bar{h}f\ ]{} & q \end{array} \qquad \text{where } h \in F$$

corresponds to
$$\begin{array}{ccccc} n & \xrightarrow{\ f\ } & p & \longrightarrow & 1 \\ \downarrow & & \downarrow{\scriptstyle \bar{h}} & & \downarrow \\ n & \xrightarrow[\ \bar{h}f\ ]{} & q & \longrightarrow & 1 \end{array} \quad \text{in } H;$$

$$M(\hat{\xi}) = \begin{array}{rcl} p(n_{\xi 1}, \ldots, n_{\xi p}) = n & \xrightarrow{\ f\ } & p \\ {\scriptstyle \xi(1, \ldots, 1)}\downarrow & & \downarrow{\scriptstyle \xi} \\ p(n_1, \ldots, n_p) \ = n & \longrightarrow & p \end{array} \ ;$$

$$M(\phi\xi) = \begin{array}{ccc} n & \longrightarrow & 1 \\ {\scriptstyle \xi}\downarrow & & \downarrow \\ n & \longrightarrow & 1 \end{array} \ ; \text{ and}$$

$M(f \otimes g) = Mf \otimes Mg.$

It can easily be seen that M preserves the relations (1), (2), and (3) on the generators of the morphisms of F'_B. Thus M is a functor.

We define a functor $N\colon H' \to F'_B$. We shall see later that $MN = 1$, $NM = 1$. Let

$N(f\colon n \to p) = \phi(f^{-1}1) \otimes \ldots \otimes \phi(f^{-1}p);$

$$N(\ \ (3)\ \ ) = \begin{array}{c} \phi(f^{-1}\xi 1) \otimes \ldots \otimes \phi(f^{-1}\xi p) \\ \downarrow{\scriptstyle \hat{\xi}} \\ \phi(f^{-1}1) \otimes \ldots \otimes \phi(f^{-1}p); \end{array}$$

N((4)) = the morphism in F corresponding to

$$\begin{array}{ccccc} n & \xrightarrow{f} & p & \longrightarrow & 1 \\ {\scriptstyle 1}\downarrow & & \downarrow{\scriptstyle h} & & \downarrow \\ n & \xrightarrow[hf]{} & q & \longrightarrow & 1 \end{array}$$ in H; and

$$N(\ (5)\ ) = \begin{array}{c} \phi a_1 \otimes \ldots \otimes \phi a_p \\ \downarrow \phi\xi_1 \otimes \ldots \otimes \phi\xi_p \\ \phi a_1 \otimes \ldots \otimes \phi a_p \end{array} .$$

Since any morphism (1) of H' can be written in the form (2), we define $N(\ (1)\ ) = N(\xi(\eta^f)^{-1},1).N(1,k).\ N(\eta^f,\eta)$. It is necessary to show that N is well defined.

We suppose that h may be written $k\eta'$. We have seen that $\eta^* = \eta'\eta^{-1}$ can be written as

$$1_q(\eta_1, \ldots, \eta_q)\colon q(k^{-1}1, \ldots, k^{-1}q) \to q(k^{-1}1, \ldots, k^{-1}q).$$

Denote $\eta' f(\eta'^f)^{-1}$ by g. Then

$$N\left(\begin{array}{ccc} n & \xrightarrow{g} & p \\ {\scriptstyle (\eta^*)^g}\downarrow & & \downarrow{\scriptstyle \eta^*} \\ n & \longrightarrow & p \\ {\scriptstyle 1}\downarrow & & \downarrow{\scriptstyle k} \\ n & \xrightarrow[d]{} & q \end{array}\right)$$ (where d is the map which makes the diagram commute)

$$= \begin{array}{l} \phi(g^{-1}\eta^*1) \otimes \ldots \otimes \phi(g^{-1}\eta^*p) \\ \quad\downarrow \hat{\eta^*} = \eta_1 \otimes \ldots \otimes \eta_q \\ \phi(g^{-1}1) \otimes \ldots \otimes \phi(g^{-1}p) \\ \quad\downarrow \bar{k} \\ \phi(d^{-1}1) \otimes \ldots \otimes \phi(d^{-1}q) \end{array} .$$

(where $\bar{k} = N(1,k)$)

$$
\begin{array}{ll}
 & \phi(g^{-1}\eta^{*}1) \otimes \ldots \otimes \phi(g'\eta^{*}p) \\
= & \quad \big\downarrow \bar{k} \\
 & \phi(d^{-1}1) \otimes \ldots \otimes \phi(d^{-1}q) \\
 & \quad \big\downarrow \phi(\eta_1^{*}(1, \ldots, 1)) \otimes \ldots \otimes \phi(\eta_q^{*}(1, \ldots, 1)) \\
 & \phi(d^{-1}1) \otimes \ldots \otimes \phi(d^{-1}q)
\end{array}
$$

(by relations (1) and (3d))

$$
= N \left(\begin{array}{ccc}
n & \xrightarrow{\ g\ } & p \\
{\scriptstyle 1}\big\downarrow & & \big\downarrow{\scriptstyle k} \\
n & \xrightarrow{\ kg\ } & q \\
{\scriptstyle (\eta^{*})^{g}}\big\downarrow & & \big\downarrow{\scriptstyle 1} \\
n & \xrightarrow[\ d\ ]{} & q
\end{array} \right)
$$

Thus $N(\xi(\eta'^{f})^{-1}, 1).\ \ N(1,k).\ \ N(\eta'^{f},\eta')$

$= N(\xi(\eta'^{f})^{-1}, 1).\ \ N(1,k).\ \ N((\eta^{*})^{g},\eta^{*}).\ \ N(\eta^{f},\eta)$

$= N(\xi(\eta'^{f})^{-1}, 1).\ \ N((\eta^{*})^{g}, 1).\ \ N(1,k).\ \ N(\eta^{f},\eta)$

$= N(\xi(\eta^{f})^{-1}1).\ \ N(1,k).\ \ N(\eta^{f},\eta)$

So N is well defined.

To show N is a functor it now suffices to show that N preserves composition. We use the following lemmas.

<u>Lemma 2.2</u>. <u>If</u> α <u>and</u> α' <u>are composable generators of type</u> (*) <u>of the morphisms of</u> H', <u>then</u> $\alpha'\alpha$ <u>is of type</u> (*), and $N(\alpha'\alpha) = N(\alpha')\, N(\alpha)$, <u>where</u> * <u>is</u> (1) C, (2) D, (3) E.

<u>Proof</u>: (1) Let $g = \xi f(\xi^{f})^{-1}$. Then

$$
N \left(\begin{array}{ccc}
n & \xrightarrow{\ g\ } & p \\
{\scriptstyle \eta^{g}}\big\downarrow & & \big\downarrow{\scriptstyle \eta} \\
n & \longrightarrow & p
\end{array} \right)
\ .\
N \left(\begin{array}{ccc}
n & \xrightarrow{\ f\ } & p \\
{\scriptstyle \xi^{f}}\big\downarrow & & \big\downarrow{\scriptstyle \xi} \\
n & \xrightarrow[\ g\ ]{} & p
\end{array} \right)
= N \left(\begin{array}{ccc}
n & \xrightarrow{\ f\ } & p \\
{\scriptstyle (\eta\xi)^{f}}\big\downarrow & & \big\downarrow{\scriptstyle \eta\xi} \\
n & \longrightarrow & p
\end{array} \right)
$$

by relation (3b).

(2)
$$N\begin{pmatrix} n & \xrightarrow{hf} & q \\ 1\downarrow & & \downarrow k \\ n & \xrightarrow[khf]{} & r \end{pmatrix} . N\begin{pmatrix} n & \xrightarrow{f} & p \\ 1\downarrow & & \downarrow h \\ n & \xrightarrow[hf]{} & q \end{pmatrix} = N\begin{pmatrix} n & \xrightarrow{f} & p \\ 1\downarrow & & \downarrow kh \\ n & \xrightarrow[khf]{} & r \end{pmatrix}$$

by relation (3a).

(3)
$$N\begin{pmatrix} n & \xrightarrow{f} & p \\ \xi\downarrow & & \downarrow 1 \\ n & \xrightarrow[f]{} & p \end{pmatrix} . N\begin{pmatrix} n & \xrightarrow{f} & p \\ \eta\downarrow & & \downarrow 1 \\ n & \xrightarrow[f]{} & p \end{pmatrix} = N\begin{pmatrix} n & \xrightarrow{f} & p \\ \xi\eta\downarrow & & \downarrow 1 \\ n & \xrightarrow[f]{} & p \end{pmatrix}$$

by relation (3c). QED.

Lemma 2.3: In Lemma 2.1, $N(\alpha)N(\beta) = N(\alpha')N(\beta')$

Proof (1) By relations (3a) and (3d).

(2) By relation (2b).

(3) By relations (1), (2a) and (3a). QED.

Let $\alpha = \varepsilon\ \delta\ \gamma$ and $\alpha' = \varepsilon'\ \delta'\ \gamma'$ be two composable morphisms of H'; where γ and γ' are of type (C), δ and δ' are of type (D), and ε and ε' are of type (E). Then

$$\begin{aligned} & \alpha'\ \alpha \\ = \ & \varepsilon'\ \delta'\ \gamma'\ \varepsilon\ \delta\ \gamma \\ = \ & \varepsilon'\ \delta'\ \varepsilon_1\ \gamma_1\ \delta\ \gamma && \text{by Lemma 2.1(2)} \\ = \ & \varepsilon'\ \varepsilon_2\ \delta_1\ \gamma_1 \delta\ \gamma && \text{by Lemma 2.1(3)} \\ = \ & \varepsilon'\ \varepsilon_2\ \delta_1\ \delta_2\ \gamma_2\ \gamma && \text{by Lemma 2.1 (1)} \end{aligned}$$

where γ_1 and γ_2 are of type (C); δ_1 and δ_2 are of type (D), ε_1 and ε_2 are of type (E). But

$$\begin{aligned} & N(\alpha')\ N(\alpha) \\ = \ & N(\varepsilon'\ \delta'\ \gamma')\ N(\varepsilon\ \delta\ \gamma) \\ = \ & N(\varepsilon')\ N(\delta')\ N(\gamma')\ N(\varepsilon)\ N(\delta)\ N(\gamma) \end{aligned}$$

$= \quad N(\varepsilon') \, N(\varepsilon_2) \, N(\delta_1) \, N(\delta_2) \, N(\gamma_2) \, N(\gamma)$ by Lemma 2.3

$= \quad N(\varepsilon'\varepsilon_2) \, N(\delta_1\delta_2) \, N(\gamma_2\gamma)$ by Lemma 2.2

$= \quad N(\alpha'\alpha)$

We have thus shown that $N\colon H' \to F'_B$ is a functor. It is easy to check that $NM = 1_{F'_B}$ and $MN = 1_{H'}$. Thus we have proved:

Theorem 2.4. H' is isomorphic to F'_B.

We now consider the whole club, F^*, i.e. we now admit the type B object $\underline{1}'$. Clearly F^*_A is isomorphic to $\underline{P}$.

The objects of F^*_B are $C_1 \otimes \ldots \otimes C_m$ where C_i is either $\underline{1}'$ or ϕn_i. (For convenience we write $\underline{1}'$ as $\underline{1}$). The generators of the morphisms of F^*_B are:

(1) Expanded instances of morphisms of F'_B; and

(2) $\hat{\xi}\colon C_{\xi 1} \otimes \ldots \otimes C_{\xi m} \to C_1 \otimes \ldots \otimes C_m$ where $\xi \in S(m)$.

Relations are:

(1) Functorial, as relation (1) for F';

(2) Naturality, as relation (2b) for F'; and

(3) Given:

(a) If $h_1 = 1\otimes k_1 \otimes 1\colon C\otimes D_1 \otimes E \to C\otimes D_2 \otimes E$ and $h_2 = 1\otimes k_2\otimes 1\colon C\otimes D_2\otimes E \to C\otimes D_3\otimes E$ are expanded instances of $k_1, k_2 \in F'_B$, then $h_2 h_1 = 1\otimes k_2 k_1 \otimes 1\colon A\otimes B_1\otimes C \to A\otimes B_3\otimes C$; and

(b) $\hat{\xi}\hat{\eta} = \widehat{\xi\eta}$.

Let H^* be the following category. An object of H^*, $(f\colon n \to p, u, \alpha)$ consists of an object $f\colon n \to p$ of H', a non-negative integer u, and a (p,u) - shuffle α. (i.e. $\alpha \in S(p+u)$. If $i<j\le p$ then $\alpha i<\alpha j$. If $p<i<j$ then $\alpha i<\alpha j$). A morphism of H^* from $(f: n \to p,u,\alpha)$ to $(g: n \to q,u,\beta)$ is (ξ,h,θ), where $(\xi,h): (f: n \to p) \to (g: n \to q)$ is a morphism of H', and $\theta \in S(u)$. The composite of (ξ',h',θ') and (ξ,h,θ) is $(\xi'\xi, h'h, \theta'\theta)$.

<u>Theorem 2.5</u>; <u>F^*_B is isomorphic to H^*.</u>

<u>Proof.</u> We shall define functors $M^*: F^*_B \to H^*$ and $N^*: H^* \to F^*_B$, such that $M^*N^* = 1_{H^*}$, and $N^*M^* = 1_{F^*_B}$.

Let M^* applied to

$$\phi n_{11} \otimes \ldots \otimes \phi n_{1p_1} \otimes \underline{1} \otimes n_{21} \otimes \ldots \otimes \phi n_{2p_2} \otimes \underline{1} \otimes \ldots \otimes \underline{1} \otimes \phi n_{u+1,1} \ldots$$
$$\ldots \otimes \phi n_{u+1,p_{u+1}}$$

be $(f: n \to p, u, \alpha)$, where $f: n \to p$ is

$$M(\phi n_{11} \otimes \ldots \otimes \phi n_{1p_1} \otimes \ldots \otimes \phi n_{u+1,1} \otimes \ldots \otimes \phi n_{u+1,p_{u+1}})$$

Thus $n = \sum_{i=1}^{u+1} \sum_{j=1}^{p_1} n_{ij}$ and $p = \sum_{i=1}^{u+1} p_i$; u is the number of occurrences of $\underline{1}$; and α is the permutation sending

$$(1,2,\ldots,p_1,p_1+1,\ldots,p_1+p_2,p_1+p_2+1, \ldots, p_1+p_2 + \ldots + p_{u+1},p+1,p+2,\ldots \ldots p+u)$$

to

$$(1,2, \ldots, p_1,p+1,p_1+1,\ldots,p_1+p_2,p+2,p_1+p_2+1, \ldots, p+u,p_1+p_2+\ldots+p_u+1, \ldots, p_1 + p_2 + \ldots + p_{u+1})$$

Let $1\otimes k\otimes 1: C_1\otimes D\otimes C_2 \to C_1\otimes D'\otimes C_2$ be an expanded instance of $k \in F'_B$. Suppose $M^*(C_i) = (f_i: n_i \to p_i, u_i, \alpha_i)$ for $i = 1,2$; and suppose $M(k) = (\xi,h)$. Then $M^*(1\otimes k\otimes 1) = (\xi',h',1)$ where (ξ',h') is:

$$\begin{array}{c}(f_1: n_1 \to p_1) \otimes M(D) \otimes (f_2: n_2 \to p_2) \\ \Big\downarrow 1\otimes(\xi,h) \otimes 1 \\ (f_1: n_1 \to p_1) \otimes M(D') \otimes (f_2: n_2 \to p_2)\end{array}$$

Suppose $\hat{\xi}$: C → D where M*C = (f: n → p, u, α) and M*D = (g: n → p, u, β). The permutation:

$$2(p,u) = p+u \xrightarrow{\alpha} p+u \xrightarrow{\xi} p+u \xrightarrow{\beta^{-1}} p+u = 2(p,u)$$

may be written as $2(\xi_1,\xi_2)$ for $\xi_1 \in S(p)$, $\xi_2 \in S(u)$. Let $M^*(\hat{\xi})$ be:

$$\left(\begin{array}{ccc} n & \xrightarrow{f} & p \\ \xi_1^{f}\Big\downarrow & & \Big\downarrow \xi_1 \\ n & \xrightarrow[g]{} & p \end{array}, \quad \begin{array}{c} u \\ \Big\downarrow \xi_2 \\ u \end{array}\right)$$

It can be verified that M* preserves the relations (1), (2), (3a) and (3b), so is a functor from $F^{\#}_{\underline{B}}$ to H^*.

We now define $N^*: H^* \to F^{\#}_{\underline{B}}$. We require N*(f: n → p, u, α). We define this to be N(f: n → p) with 1' inserted u times at the positions α(p+1), α(p+2), ... α(p+u). It is easily seen that M*N* = 1 and N*M* = 1 on objects.

There is an isomorphism $\hat{\alpha}$ in $F^{\#}_{\underline{B}}$:

$\hat{\alpha}$: N*(f: n → p, u, α) → N(f: n → p) ⊗ $\underline{1}^u$

where $\underline{1}^u$ is $\underline{1}\otimes\underline{1}\otimes \dots \otimes\underline{1}$, where $\underline{1}$ is mentioned u times. If (ξ,h,θ): (f: n → p, u, α) → (g: n → q, u, β) define N*(ξ,h,θ) to be the composite:

$$\begin{array}{c} N^*(f\colon n \to p,\ u,\ \alpha) \\ \big\downarrow \hat{\alpha} \\ N(f\colon n \to p) \otimes \underline{1}^u \\ \big\downarrow N(\xi,h) \otimes \hat{\theta} \\ N(g\colon n \to q) \otimes \underline{1}^u \\ \big\downarrow \hat{\beta}^{-1} \\ N^*(g\colon n \to q,\ u,\ \beta) \end{array}$$

It can be verified that $M^*N^* = 1$ and $N^*M^* = 1$. QED.

§3 Determination of the club in the non-strict monoidal case

We shall treat the symmetric case. The non-symmetric case follows a similar argument. We shall show that the club in the non-strict case is equivalent to the club for the strict case.

The type A objects for the club E are I, $\underline{1}$, and $T\otimes S$ where T and S are type A objects. The type B objects are I', $\underline{1}'$, $T\otimes' T'$ and ϕS, where T and T' are type B, and S is type A.

The type A morphisms of E are generated by expansions of the following instances:

(1) $a\colon (T\otimes S)\otimes R \to T\otimes(S\otimes R)$ and a^{-1}, $T\otimes(S\otimes R) \to (T\otimes S)\otimes R$;

(2) $\ell\colon I\otimes T \to T$, $r\colon T\otimes I \to T$, $\ell^{-1}\colon T \to I\otimes T$, $r^{-1}\colon T \to T\otimes I$;

(3) $c\colon T\otimes S \to S\otimes T$

for type A objects T,S,R.

The type B morphisms of E are generated by expansions of:

(1) a', a'^{-1};

(2) ℓ', r', ℓ'^{-1}, r'^{-1};

(3) c';

(4) $\phi f: \phi T \to \phi S$, where $f: T \to S$ is a type A generator;

(5) $\tilde{\phi}$, ϕ^o.

Type A morphisms satisfy the following relations:

(1) Functoriality of $\otimes$;

(2) (a) Naturality of a;

(b) Naturality of r (naturality of $r^{-1}, l, l^{-1}, a^{-1}$ will follow from relations (2a), (2b), (3a) and (3c));

(c) Naturality of c;

(3) (a) $aa^{-1} = 1, a^{-1}a = 1, \ell\ell^{-1} = 1, \ell^{-1}\ell = 1, rr^{-1} = 1, r^{-1}r = 1$;

(b) The diagrams MC1 - 7 of [1]; and

(c)
$$\begin{array}{ccc} I\otimes T & \xrightarrow{\ c\ } & T\otimes I \\ & {}_{\ell}\searrow \quad \swarrow_{r} & \\ & T & \end{array}$$

The type B morphisms satisfy the following relations:

(1) (a) Functoriality of $\otimes'$;

(b) Functoriality of ϕ, i.e. $\phi gf = \phi g . \phi f$;

(2) (a) Naturality of a', r', c' (and consequently ℓ', a'^{-1}, r'^{-1}, ℓ'^{-1});

(b) Naturality of $\tilde{\phi}$;

(3) (a) $a' a'^{-1} = 1, \ldots,$ as in (3a) for type A morphisms;

(b) As (3b) for type A morphisms;

(c) As (3c) for type A morphisms;

(d) The diagrams MF2, MF3 and MF4 of [1].

Theorem 3.1 We describe a surjective functor $J: \mathcal{E} \to \mathcal{F}^*$ which is full and faithful, i.e. $\mathcal{E}(T,S) \cong \mathcal{F}^*(JT,JS)$. Thus $\mathcal{E}$ is equivalent to $\mathcal{F}^*$.

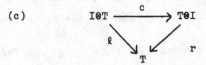

<u>Proof</u>: For each object T of E, we describe an object K(T) and an isomorphism $\eta(T)\colon T \to K(T)$ in E. $\eta(T)$ will not involve c, c', $\tilde{\phi}$ or ϕ^{o}. Let $K_0 = I$, $K_1 = \underline{1}$, $K_n = K_{n-1}\otimes\underline{1}$. Let $K'_0 = I'$, $K'_1 = \underline{1}'$, $K'_n = K'_{n-1}\otimes'\underline{1}'$. Define $\theta_{n,m}\colon K_n\otimes K_m \to K_{n+m}$ by:

$$\theta_{n,0} = r\colon K_n\otimes I \to K_n;$$

$$\theta_{n,1} = 1\colon K_n\otimes\underline{1} \to K_{n+1};$$

$\theta_{n,m}$ is the composite:

$$K_n\otimes K_m \xrightarrow{a^{-1}} (K_n\otimes K_{m-1})\otimes\underline{1} \xrightarrow{\theta_{n,m-1}\otimes 1} K_{n+m-1}\otimes\underline{1} = K_{n+m}$$

Define $\theta'_{n,m}\colon K'_n\otimes' K'_m \to K'_{n+m}$ analogously. $\theta_{n,m}$ and $\theta'_{n,m}$ are isomorphisms not involving c, c', $\tilde{\phi}$ and ϕ^{o}.

Let $K(I) = I$, $K(\underline{1}) = \underline{1}$, and $\eta(I) = 1$, $\eta(\underline{1}) = 1$. If $K(T) = K_m$ and $K(S) = K_n$, then let $K(T\otimes S) = K_{m+n}$ and $(T\otimes S)$ be the composite:

$$T\otimes S \xrightarrow{\eta T\otimes\eta S} K_m\otimes K_n \xrightarrow{\theta_{m,n}} K_{m+n}.$$

Let $K(I') = I'$, $K(\underline{1}') = \underline{1}'$, and $\eta I' = 1$, $\eta\underline{1}' = 1$. If $KT = K'_m(A_1, \ldots, A_m)$ and $KS = K'_n(B_1, \ldots, B_n)$, let $K(T\otimes S)$ be $K'_{m+n}(A_1, \ldots, A_m, B_1, \ldots, B_n)$ and $\eta(T\otimes' S)$ be the composite:

$$\begin{array}{c} T\otimes' S \\ \big\downarrow \eta T\otimes'\eta S \\ K'_m(A_1, \ldots, A_m)\otimes' K'_n(B_1, \ldots, B_n) \\ \big\downarrow \theta'_{m,n}(A_1, \ldots, A_m, B_1, \ldots, B_n) \\ K'_{m+n}(A_1, \ldots, A_m, B_1, \ldots, B_n). \end{array}$$

If R is a type A object, and $KR = K_p$, then let $K(\phi R)$ be $\underline{1}'\{\phi K_p\} = \phi K_p$ and $\eta(\phi R)$ be $\phi(\eta R)\colon \phi R \to \phi K_p$.

Define L as a partial function from the objects of E to the objects of F^{*} given by:

$L(K_n) = n;$

$L(I') = \emptyset;$

$L(\underline{1}') = \underline{1}';$

$L(\phi K_n) = \phi n;$

$L(K_n'(A_1, \ldots, A_n)) = LA_1 \otimes' \ldots \otimes' LA_n$ when

LA_i is defined. When T is an object of E, define JT as L(KT). J is certainly surjective on objects, and L is injective.

Let $f: T \to S$ be a morphism in E. We may write f as $f_n f_{n-1} \ldots f_1$ where each f_i is either an expansion of an instance of a, a^{-1}, r, r^{-1}, ℓ, ℓ^{-1}, a', a'^{-1}, r', r'^{-1}, ℓ', or ℓ'^{-1}; or is of the form:

$$(1 \otimes g_i) \otimes 1: (P \otimes R) \otimes Q \to (P \otimes R') \otimes Q, \text{ or}$$

$$(1 \otimes' g_i) \otimes' 1: (P \otimes' R) \otimes' Q \to (P \otimes' R') \otimes' Q$$

where g_i is an instance of c, f', $\tilde{\phi}$, ϕ^o or ϕf. Define $J(f) = J(f_n)\, J(f_{n-1}) \ldots J(f_1)$. If F_i is of the first type, let $J(f_i) = 1$. Let $J((1 \otimes g_i) \otimes 1)$ be

$$1 \otimes Jg_i \otimes 1: JP \otimes JR \otimes JQ \to JP \otimes JR' \otimes JQ$$

and similarly for $J((1 \otimes' g_i) \otimes' 1)$; where Jg_i is defined below. If g_i is $c: R_1 \otimes R_2 \to R_2 \otimes R_1$, and $J(R_j) = r_j$, let Jg_i be $\hat{\tau}(1,1): 2(r_1, r_2) \to 2(r_2, r_1)$, where τ is the non-identity element of S(2). If g_i is $c': R_3 \otimes' R_4 \to R_4 \otimes' R_3$, let Jg_i be $\tau(1,1): 2(JR_3, JR_4) \to 2(JR_4, JR_3)$. If g_i is $\phi h: \phi R_5 \to \phi R_6$, let Jg_i be $\phi(Jh): \phi(JR_5) \to \phi(JR_6)$. If g_i is $\tilde{\phi}$ or ϕ^o, let Jg_i be $\tilde{\phi}$ or ϕ^o, respectively. Since J preserves the relations on E, J is a functor.

Suppose $f: LS \to LS'$ is an instance in F^*. We define $J^*f: S \to S'$ as a morphism of E. If f is $\xi: n \to n$, let J^*f be a composite of instances of a, a^{-1}, and c with graph ξ (unique by [6]). J^*f is defined similarly if f is $\hat{\xi}$, or $\phi f'$. If f is $\tilde{\phi}$ or ϕ^o, then so is J^*f.

Suppose f is $1\otimes' f'\otimes' 1.\ LT\otimes' LS\otimes' LR \to LT\otimes' LS'\otimes' LR$, an expansion of an instance f' in E. (The $\otimes$ case is analogous). Let J^*f be the composite:

$$\begin{array}{l} K((T\otimes' S)\otimes' R) \\ \quad\big\downarrow (\eta((T\otimes' S)\otimes' R))^{-1} \\ (T\otimes' S)\otimes' R \\ \quad\big\downarrow (1\otimes' J^*f')\otimes' 1 \\ (T\otimes' S')\otimes' R \\ \quad\big\downarrow \eta((T\otimes' S')\otimes' R) \\ K((T\otimes' S')\otimes' R) \end{array}$$

Suppose $g = 1\otimes' g_1\otimes' 1\colon (T_1\otimes' S_1)\otimes' R_1 \to (T_1\otimes' S_2)\otimes' R_1$ is a morphism of F^*, such that $KT_1 = T$, $KR_1 = R$, $KS_1 = S$, $KS_2 = S'$, and $Jg_1 = f'$. Then the composite

$$\begin{array}{l} K((T\otimes' S)\otimes' R) \\ \quad\big\downarrow (\eta((T_1\otimes' S_1)\otimes' R_1))^{-1} \\ (T_1\otimes' S_1)\otimes' R_1 \\ \quad\big\downarrow (1\otimes' g_1)\otimes' 1 \\ (T_1\otimes' S_2)\otimes' R_1 \\ \quad\big\downarrow \eta((T_1\otimes' S_2)\otimes' R_1) \\ K((T\otimes' S')\otimes' R) \end{array}$$

equals J^*f, by the naturality of $a, \ell, r, a', \ell', r'$ and their inverses, and the coherence of symmetric monoidal categories.

If $f\colon T \to S$ in E is written $f_n \ldots f_2 f_1$ where f_i is an expansion of an instance, then let J^*f be $J^*f_n \ldots J^*f_2 . J^*f_1$. The relations in F^* are preserved by J^*, so $J^*\colon F^* \to E$ is a functor.

We now describe the bijection between $E(T,S)$ and $F^*(JT,JS)$.

There is a correspondence between f: $T \to S$ and Jf: $JT \to JS$, and between g: $JT \to JS$ in F^* and the composite :

$$T \xrightarrow{\eta T} KT \xrightarrow{J^*g} KS \xrightarrow{(\eta S)^{-1}} S.$$

Thus $E(T,S)$ is isomorphic to $F^*(JT,JS)$. QED.

§4 Coherence for a closed functor

We describe N, the free model on $A = \{A,B\}$ for a closed functor between two categories.

The objects of N of type A are $\underline{1}$, I, $T \otimes S$ and $[T,S]$ where T and S are objects of type A. The objects of type B are $\underline{1}'$, I', $T \otimes' S$, $[T,S]'$ and ϕR, where T and S are of type B, and R is of type A.

Morphisms of type A are generated by expanded instances of a, a^{-1}, b, b^{-1}, c, d, e; where b: $T \otimes I \to T$, d: $T \to [S, T \otimes S]$ and e: $[T,S] \otimes T \to S$. Type B morphisms are generated by expanded instances of a', a'^{-1}, b', b'^{-1}, c', d', e', ϕf, $\tilde{\phi}$ and ϕ^o; where f is a type A generator.

Type A morphism satisfy the following relations:

(1) Functoriality of $\otimes$; and [,]:

$$\begin{array}{ccc} [T',S] & \xrightarrow{[f,1]} & [T,S] \\ {\scriptstyle [1,g]}\downarrow & & \downarrow{\scriptstyle [1,g]} \\ [T',S'] & \xrightarrow[{[f,1]}]{} & [T,S'] \end{array}$$

(2) Naturality of a, a^{-1}, b, b^{-1}, c; and d and e:

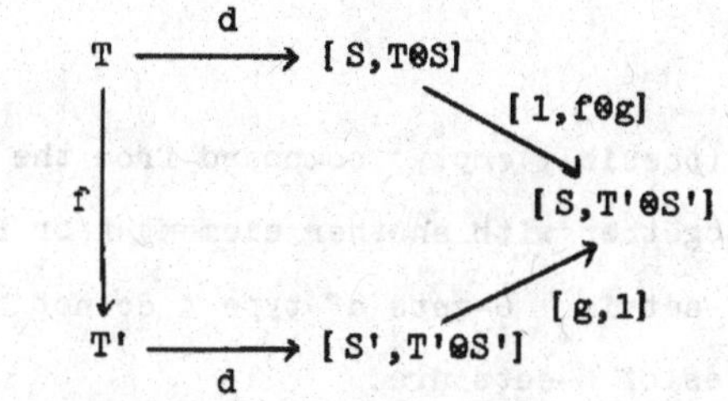

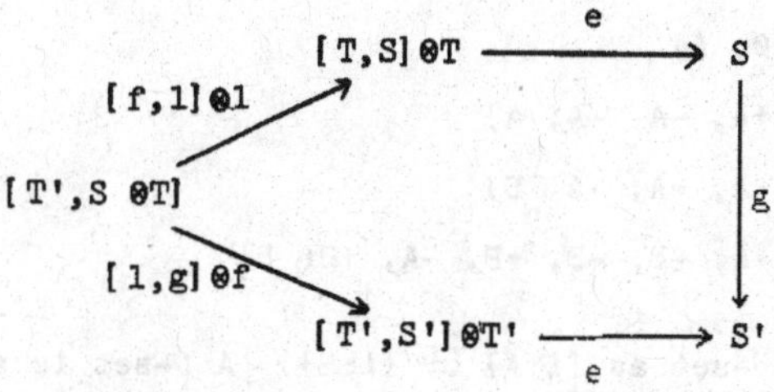

(3) $aa^{-1} = 1$, $a^{-1}a = 1$, $bb^{-1} = 1$, $b^{-1}b = 1$, and the diagrams C1-C6 of [5].

Type B morphisms satisfy the following relations:

(1) Functoriality of ⊗', [,]', and ϕ:

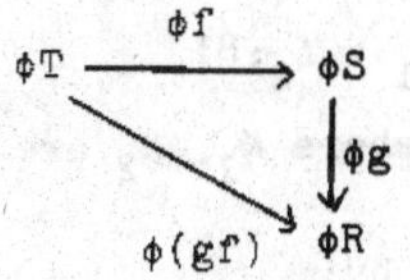

(2) Naturality of $a', a'^{-1}, b', b'^{-1}, c', d', e'$ and $\tilde{\phi}$;

(3)(a) The relations that a' and b' are isomorphisms;

(b) The diagrams C1-C6 of [5] for a', ..., e';

(c) The relations that ϕ is a closed functor, i.e. diagrams MF2, MF3 and MF4 of [1].

We now describe a precategory G with objects G-sets and morphisms graphs, and a precategory D with objects D-sets and morphisms D-graphs. G-graphs may be of type A or type B. The domain and the codomain of a graph of type A(respectively type B)are also of type A (respectively B).

A G-set is a list (possibly empty) composed from the four elements +A, -A, +B, -B; together with another element, A or B, which specifies which type the G-set is. G-sets of type A do not have +B or -B in the list. Examples of G-sets are:

$$\{\emptyset;\ A\}$$
$$\{+A,\ -A,\ -A;\ A\}$$
$$\{+A,\ -A,\ -A;\ B\}$$
$$\{+A,\ -B,\ -B,\ +B,\ -A,\ +B;\ B\}$$

We sometimes write the G-set as $\{L;A\}$ or $\{L;B\}$. A D-set is a list (possibly empty) composed from the two elements + and -. If L_1 and L_2 are lists, define $L_1 \vee L_2$ as the list comprising L_1 and L_2, with L_1 preceding L_2. Let $-L$ be the list obtained from by changing the sign of each of the elements of L. Define $\otimes$ on G-sets and D-sets by:

$$\{L_1;A\} \otimes \{L_2;A\} = \{L_1 \vee L_2;A\}$$
$$\{L_1;B\} \otimes \{L_2;B\} = \{L_1 \vee L_2;B\}$$
$$K_1 \otimes K_2 = K_1 \vee K_2, \text{ where } K_1,\ K_2 \text{ are } D\text{-sets.}$$

Define $[\,,]$ on G-sets and D-sets by:

$$[\{L_1;A\},\ \{L_2;A\}] = \{(-L_1) \vee L_2;A\}$$
$$[\{L_1;B\},\ \{L_2;B\ ] \quad \{(-L_1) \vee L_2;B\}$$
$$[K_1,\ K_2] = (-K_1) \vee K_2$$

If μ is a G-set, let μ_A^+ be the set of +A elements of the list of μ. Define μ_A^-, μ_B^+, μ_B^- similarly. If μ is a D-set, let μ^+ be the set of + elements of μ. Similarly define μ^-. A graph $f\colon \mu \to \nu$ consists of:

(1) a bijection from $\mu_A^+ \cup \nu_A^-$ to $\mu_A^- \cup \nu_A^+$; and

(2) a bijection from $\mu_B^+ \cup \nu_B^-$ to $\mu_B^- \cup \nu_B^+$.

A $\mathcal{D}$-graph $f: \mu \to \nu$ consists of a function from $\mu^+ \cup \nu^-$ to $\mu^- \cup \nu^+$.

Suppose $f: \mu \to \nu$ and $g: \nu \to \pi$ are either both graphs or both $\mathcal{D}$-graphs. Then g and f are said to be <u>incompatible</u> if there is a subset

$$\nu_1, \nu_2, \ldots, \nu_{2n} \qquad n \geq 1$$

of the elements of the list of ν, such that f maps ν_{2i-1} to ν_{2i} $(i = 1, 2, \ldots, n)$; and g maps ν_{2i} to ν_{2i+1} $(i = 1, \ldots, n-1)$, and ν_{2n} to ν_1. Otherwise we say that g and f are compatible. If g and f are compatible we define $gf: \mu \to \pi$. Suppose $\mu_0 \varepsilon \mu_A^+ \cup \mu_B^+$ for G, or $\mu_0 \varepsilon \mu^+$ for $\mathcal{D}$. Consider the sequence:

$$\mu_0, \nu_1, \nu_2, \ldots, \nu_r, \alpha$$

where $\nu_i \ \varepsilon \ L(\nu)$ (the list of ν), and $\alpha \ \varepsilon \ \mu^- \cup \pi^+$ such that:

(1) f maps μ_0 to ν_1, and ν_i to ν_{i+1} when i is even and $2 \leq i \leq r-1$;

(2) g maps ν_i to ν_{i+1} when i is odd and $1 \leq i \leq r-1$;

(3) If r is odd, $\alpha \ \varepsilon \ \pi^+$ and g maps ν_r to α.

If r is even, $\alpha \ \varepsilon \ \mu^-$ and f maps ν_r to α.

We define a similar sequence

$$\pi_0, \nu_1', \nu_2', \ldots, \nu_r', \beta$$

for $\pi_0 \ \varepsilon \ \pi'$. Then $\beta \varepsilon \mu^- \cup \pi^+$. The correspondences between μ_0 and α, and π_0 and β set up a function from $\mu^+ \cup \pi^-$ to $\mu^- \cup \pi^+$. In G this function consists of two bijections between $\mu_A^+ \cup \pi_A^-$ and $\mu_A^- \cup \pi_A^+$, and $\mu_B^+ \cup \pi_B^-$ and $\mu_B^- \cup \pi_B^+$. Thus we have defined gf: $\mu \to \pi$ for graphs and $\mathcal{D}$-graphs.

Theorem 4.1 Suppose f: $\mu \to \nu$, g: $\nu \to \pi$ and h: $\pi \to \rho$ are all graphs or all $\mathcal{D}$-graphs. Then g and f are compatible, and h and gf are compatible iff h and g are compatible, and hg and f are compatible.

Proof: Suppose g and f are incompatible. Then there exist

$$\nu_1, \ldots, \nu_{2n} \qquad n \geq 1$$

in $L(\nu)$, such that f maps ν_{2i-1} to ν_{2i} for i = 1, ..., n; and g maps ν_{2n} to ν_1, and ν_{2i} to ν_{2i+1} for i = 1, ..., n-1. But if h and g are compatible then hg maps ν_{2n} to ν_1, and ν_{2i} to ν_{2i+1}. Thus hg and f are incompatible.

Suppose g and f are compatible, and h and gf are incompatible. Then there exist

$$\pi_1, \ldots, \pi_{2n} \qquad n \geq 1$$

in $L(\pi)$, such that gf maps π_{2i-1} to π_{2i}, and h maps π_{2n} to π_1, and π_{2i} to π_{2i+1}. Since gf maps π_{2i-1} to π_{2i}, there exists a sequence (possibly empty)

$$\nu_{i,1}, \nu_{i,2}, \ldots, \nu_{i,r_i} \quad \text{(where } r_i \text{ is even)}$$

in $L(\nu)$ such that g maps π_{2i-1} to $\nu_{i,1}$, ν_{i,r_i} to π_{2i}, and $\nu_{i,j}$ to $\nu_{i,j+1}$ when j is even; and f maps $\nu_{i,j}$ to $\nu_{i,j+1}$ when j is odd. If $r_i = 0$ for all i, then h and g are incompatible. Suppose r_i is not always 0. Consider:

$$\nu_{1,1}, \nu_{1,2}, \ldots, \nu_{1,r_1}, \nu_{2,1}, \ldots, \nu_{2,r_2}, \ldots, \nu_{n,1}, \ldots, \nu_{n,r_n}.$$

When j is even, hg maps $\nu_{i,j}$ to $\nu_{i,j+1}$. If ν_{i,r_i} and $\nu_{j,1}$ are consecutive terms in the sequence (including the last and first terms), then hg maps ν_{i,r_i} to $\nu_{j,1}$. Thus hg and f are incompatible. QED.

<u>Theorem 4.2</u> <u>Suppose</u> $f: \mu \to \nu$, $g: \nu \to \pi$ <u>and</u> $h: \pi \to \rho$ <u>are all graphs or all</u> $\mathcal{D}$<u>-graphs</u>. <u>Suppose that</u> g <u>and</u> f <u>are compatible</u>, <u>and</u> h <u>and</u> gf <u>are compatible.</u> <u>Then</u> $h(gf) = (hg)f$.

<u>Proof</u>: Suppose $\mu_0 \in \mu^+$. If $(h(gf))\mu_0 = \alpha$, then there is a sequence

$$\mu_0, \pi_1, \ldots, \pi_s, \alpha,$$

where $\pi_i \in L(\pi)$ and $\alpha \in \mu^- \cup \rho^+$; such that gf maps μ_0 to π_1, π_{2i} to π_{2i+1}, and π_s to α if s is even, and h maps π_{2i-1} to π_{2i}, and π_s to α if s is odd. We have the following sequences in $L(\nu)$.

(1) $\nu_{0,1}, \ldots, \nu_{0,r_0}$ for r_0 odd,

where f maps μ_0 to $\nu_{0,1}$, and $\nu_{0,i}$ to $\nu_{0,i+1}$ when i is even; and g maps ν_{0,r_0} to π_1, and $\nu_{0,i}$ to $\nu_{0,i+1}$ when i is odd.

(2) $\nu_{i,1}, \ldots, \nu_{i,r_i}$ for r_0 even,

where f maps $\nu_{i,j}$ to $\nu_{i,j+1}$ when j is odd, and g maps π_{2i} to $\nu_{i,1}$, ν_{i,r_i} to π_{2i+1}, and $\nu_{i,j}$ to $\nu_{i,j+1}$ when j is even.

(3) If s is even:

$\nu_{s,1}, \ldots, \nu_{s,r_s}$ for r_s odd,

where f maps ν_{s,r_s} to α, and $\nu_{s,i}$ to $\nu_{s,i+1}$ for i odd; and g maps π_s to $\nu_{s,1}$, and $\nu_{s,i}$ to $\nu_{s,i+1}$ for i even. Consider the sequence

$$\mu_0, \nu_{0,1}, \ldots, \nu_{0,r_0}, \nu_{1,1}, \ldots, \nu_{1,r_1}, \ldots, \nu_{t,1}, \ldots, \nu_{t,r_t}, \nu_{s,1}, \ldots, \nu_{s,r_s}, \alpha$$

where $t = \frac{s-2}{2}$ if s is even, and $t = \frac{s-1}{2}$ if s is odd. We see that f maps μ_0 to $\nu_{0,1}$; $\nu_{0,i}$ to $\nu_{0,i+1}$ when i is even; $\nu_{i,j}$ to $\nu_{i,j+1}$ when $i = 1, \ldots, t$ and j is odd; $\nu_{s,i}$ to $\nu_{s,i+1}$ when i is even; and ν_{s,r_s} to α. Also hg maps $\nu_{0,i}$ to $\nu_{0,i+1}$ when i is odd; $\nu_{i,j}$ to $\nu_{i,j+1}$

when $i = 1, \ldots, t$ and j is even; $\nu_{s,i}$ to $\nu_{s,i+1}$ when i is odd; ν_{i,r_i} to $\nu_{j,1}$ when ν_{i,r_i} and $\nu_{j,1}$ are consecutive terms in the sequence; and ν_{t,r_t} to α if s is odd. Thus $\alpha = ((hg\ )f)\mu_0$. Similarly $(h(gf))\rho_0 = ((hg)f)\rho_0$ for $\rho_0 \varepsilon \rho^-$. Thus $h(gf) = (hg)f$. QED.

Suppose $f: \mu \to \nu$ and $g: \pi \to \rho$ are either both graphs or both $\mathcal{D}$-graphs. We define $f \otimes g: \mu x \pi \to \nu \otimes \rho$ and $[f,g]: [\nu,\pi] \to [\mu,\rho]$. Let $f \otimes g$ map α to β if either f maps α to β or g maps α to β. Let $[f,g]$ map α to β if either g maps α to β or f maps α to β.

Let us define the categories $\mathcal{G}^*$ and $\mathcal{D}^*$. The objects of $\mathcal{G}^*$ and $\mathcal{D}^*$ are the objects of $\mathcal{G}$ and $\mathcal{D}$ respectively. Let $\mathcal{G}^*(\mu,\nu) = \mathcal{G}(\mu,\nu) \cup \{*\}$ and $\mathcal{D}^*(\mu,\nu) = \mathcal{D}(\mu,\nu) \cup \{*\}$. If $f: \mu \to \nu$ and $g: \nu \to \pi$ are either both in $\mathcal{G}^*$ or both in $\mathcal{D}^*$, then their composite is:

(1) the composite in $\mathcal{G}$ or $\mathcal{D}$, respectively, if g and f are compatible;

(2) $*$, if g and f are incompatible; or

(3) $*$, if either g or f is $*$.

By Theorems 4.1 and 4.2 composition in $\mathcal{G}^*$ and $\mathcal{D}^*$ is associative. Define $\otimes$ and $[\ ,\ ]$ in $\mathcal{G}^*$ and $\mathcal{D}^*$ as in $\mathcal{G}$ and $\mathcal{D}$, except that $f \otimes g = *$ and $[f,g] = *$, if either f or g is $*$. Since there are obvious identity graphs and $\mathcal{D}$-graphs $\mathcal{G}^*$ and $\mathcal{D}^*$ are categories.

Define $\gamma(\mu,\nu): \mu\vee\nu \to \nu\vee\mu$ by letting $\gamma(\mu,\nu)(x)$ equal:

$x \varepsilon \mu^+ \subset (\nu\vee\mu)^+$, when $x \varepsilon \mu^+ \subset (\mu\vee\nu)^+$;
$x \varepsilon \nu^+ \subset (\nu\vee\mu)^+$, when $x \varepsilon \nu^+ \subset (\mu\vee\nu)^+$;
$x \varepsilon \mu^- \subset (\mu\vee\nu)^-$, when $x \varepsilon \mu^- \subset (\mu\vee\nu)^-$; and
$x \varepsilon \nu^- \subset (\mu\vee\nu)^-$, when $x \varepsilon \nu^- \subset (\mu\vee\nu)^-$.

Define $\delta(\mu,\nu): \mu \to (-\nu)\vee\mu\vee\nu = \pi$ by letting $\delta(\mu,\nu)(x)$ equal:

$x \in \mu^+ \subset \pi^+$, when $x \in \mu^+$;

$x \in \mu^-$, when $x \in \mu^- \subset \pi^-$;

$-x \in (-\nu)^+ \subset \pi^+$, when $x \in \nu^- \subset \pi^-$; and

$-x \in \nu^+ \subset \pi^+$ when $x \in (-\nu)^- \subset \pi^-$.

Define $\varepsilon(\mu,\nu)\colon (-\mu)\vee\nu\vee\mu \to \nu$ as $(\delta(\nu,\mu))^{-1}$.

We define a functor $\Gamma\colon N \to G^*$:

$\Gamma I = \{\emptyset;A\}$: $\qquad$ $\Gamma I' = \{\emptyset;B\}$;

$\Gamma \underline{\underline{1}} = \{A;A\}$; $\qquad$ $\Gamma \underline{\underline{1}}' = \{B;B\}$;

$\Gamma(T\otimes S) = \Gamma T\otimes\Gamma S$; $\qquad$ $\Gamma(T\otimes' S) = \Gamma T\otimes\Gamma S$;

$\Gamma([T,S]) = [\Gamma T,\Gamma S]$; $\qquad$ $\Gamma([T,S]') = [\Gamma T,\Gamma S]$;

If $\Gamma S = \{L;A\}$, then $\Gamma(\phi S) = \{L;B\}$;

$\Gamma a = \Gamma a^{-1} = \Gamma b = \Gamma b^{-1} = 1$;

$\Gamma(c\colon T\otimes S \to S\otimes T) = \gamma(L(T), L(S))$;

$\Gamma d = \delta$, $\Gamma e = \varepsilon$;

$\Gamma(f\otimes g) = \Gamma f\otimes\Gamma g$;

$\Gamma([f,g]) = [\Gamma f,\Gamma g]$;

$\Gamma a' = \Gamma a'^{-1} = \Gamma b' = \Gamma b'^{-1} = 1$;

$\Gamma c' = \gamma$, $\Gamma d' = \delta$, $\Gamma e' = \varepsilon$;

$\Gamma(f\otimes' g) = \Gamma f\otimes\Gamma g$;

$\Gamma([f,g]') = [\Gamma f,\Gamma g]$;

$\Gamma(\phi f) = \Gamma f$;

$\Gamma(\tilde{\phi}) = 1$, $\Gamma(\phi^o) = 1$; and

$\Gamma(gf) = \Gamma g.\ \Gamma f$.

We define a functor $\Delta\colon N \to D^*$:

$\Delta T = \{\emptyset\}$ for all objects of N_A;

$\Delta I' = \Delta\underline{\underline{1}}' = \{\emptyset\}$; $\Delta(\phi S) = \{+\}$;

$\Delta(T\otimes' S) = \Delta T\otimes\Delta S$: $\Delta([T,S]') = [\Delta T,\Delta S]$;

$\Delta f = 1\colon \{\emptyset\} \to \{\emptyset\}$ for all morphisms of N_A;

$\Delta a' = \Delta a'^{-1} = \Delta b' = \Delta b'^{-1} = 1$;

$\Delta c' = \gamma$, $\Delta d' = \delta$, $\Delta e' = \varepsilon$;

$\Delta(f \otimes' g) = \Delta f \otimes \Delta g$; $\Delta([f,g]') = [\Delta f, \Delta g]$;

$\Delta(\phi f) = 1: \{+\} \to \{+\}$;

$\Delta(\tilde{\phi})$ is the $\mathcal{D}$-graph: $\{+, +\} \to \{+\}$;

$\Delta(\phi^{o})$ is the $\mathcal{D}$-graph: $\{\emptyset\} \to \{+\}$; and

$\Delta(gf) = \Delta g . \Delta f$.

It can easily be verified that Γ and Δ preserve the relations on $\mathcal{N}$.

Let the central morphism of $\mathcal{N}_A$ be those generated by a, a^{-1}, b, b^{-1} and c. Let the central morphisms of $\mathcal{N}_B$ be those generated by a', a'^{-1}, b', b'^{-1}, c' and ϕf where f is central of type A.

If $f: A \otimes B \to C$ is a morphism of $\mathcal{N}_A$, let πf be the composite

$$A \xrightarrow{d} [B, A \otimes B] \xrightarrow{[1,f]} [B,C]$$

If $g: D \to E$, let $\langle g \rangle_F$ (usually written $\langle g \rangle$) be the composite

$$[E,F] \otimes D \xrightarrow{1 \otimes g} [E,F] \otimes E \xrightarrow{e} F$$

Define πf and $\langle g \rangle$ in the same way for type B.

Let the type A constructible morphisms be the smallest class of morphisms of $\mathcal{N}_A$ which satisfy the following conditions:

CA1: Every central morphism is in the class.

CA2: If $f: T \to S$ is in the class and if $u: T' \to T$ and $v: S \to S'$ are central then $vfu: T' \to S'$ is in the class.

CA3: If $f: T \to S$ and $g: P \to Q$ are in the class so is $f \otimes g: T \otimes P \to S \otimes Q$.

CA4: If $f: A \otimes B \to C$ is in the class so is $\pi f: A \to [B,C]$

CA5: If $f: T \to S$ and $g: P \otimes Q \to R$ are in the class so is $g(\langle f\rangle \otimes 1): ([S,P] \otimes T) \otimes Q \to R$.

Let the type B constructible morphisms be the smallest class of morphisms of N_B which satisfy the following conditions:

CB1: CA1

CB2: CA2

CB3: CA3 with $\otimes'$ replacing $\otimes$

CB4: CA4 with $\otimes'$ and $[\ ,\ ]'$ replacing $\otimes$ and $[\ ,\ ]$

CB5: CA5 with $\otimes'$ replacing $\otimes$

CB6: If $f: K_n(S_1, \ldots, S_n) \to T$ (notation as in §3)

is a type A constructible morphism, then the composite

$$K'_n(\phi S_1, \ldots, \phi S_n) \xrightarrow{\overline{\phi}_n} \phi(K_n(S_1,\ldots,S_n)) \xrightarrow{\phi f} \phi T$$

is a type B constructible morphism. We define $\overline{\phi}_0$ to be ϕ^o; $\overline{\phi}_1$ to be 1; and if $n > 1$, $\overline{\phi}_n$ to be the composite:

$$\begin{array}{c} K'_{n-1}(\phi S_1, \ldots, \phi S_{n-1}) \otimes' \phi S_n \\ \Big\downarrow \overline{\phi}_{n-1} \otimes' 1 \\ \phi(K_{n-1}(S_1, \ldots, S_{n-1})) \otimes' \phi S_n \\ \Big\downarrow \tilde{\phi} \\ \phi(K_n(S_1, \ldots, S_n)). \end{array}$$

An object T of N is <u>constant</u> if it does not involve $\underline{1}$ or $\underline{1}'$. An object T of N is <u>integral</u> if it does not involve $[\ ,\ ]$ or $[\ ,\ ]'$. An object T of N is <u>ϕ-free</u> if it does not involve ϕ. A morphism $f: T \to S$ of N is said to be <u>trivial</u> if T and S are constant ϕ-free and integral.

<u>Lemma 4.3</u>: <u>A trivial constructible morphism in N is central</u>.

Proof: As Lemma 6.1 of [5].

Proposition 4.4: *For each constructible* $h: T \to S$ *in* N, *at least one of the following is true*:

(1) h *is central*

(2) h *is of the form*

$$T \xrightarrow{x} A \otimes B \xrightarrow{f \otimes g} C \otimes D \xrightarrow{y} S$$

where f *and* g *are non-trivial.*

(3) h *is of the form*

$$T \xrightarrow{\pi f} [B,C] \xrightarrow{y} S$$

(4) h *is of the form*

$$T \xrightarrow{x} ([B,C] \otimes A) \otimes D \xrightarrow{\langle f \rangle \otimes 1} C \otimes D \xrightarrow{g} S$$

(5) h *is of the form*

$$T \xrightarrow{x} K'_n(\phi A_1, \ldots, \phi A_n) \xrightarrow{\bar{\Phi}_n} \phi(K_n(A_1, \ldots, A_n)) \xrightarrow{\phi f} \phi B \xrightarrow{y} S;$$

where f *and* g *are constructible, and* x *and* y *are central.*

Notation: For convenience we are using $\otimes$ to represent $\otimes$ or $\otimes'$ or both. Similarly for $[\ ,\ ]$, a,b,c,d,e,I,$\underline{1}$, etc. For brevity, morphisms h of the forms (2), (3), (4), (5) are said to be of type $\otimes$, of type π, of type $\langle\ \rangle$ and of type ϕ respectively.

Proof: The proof is the same as Proposition 6.2 of [5] except that

(1) Axiom CB6 is clearly satisfied and

(2) If B is a constant integral ϕ-free object, we

need to show that there is a central morphism $u(B): B \to I$. But B is generated by just $\otimes$ and I (or just $\otimes'$ and I'). If $B = I$, let $u = 1$.

If $B = C \otimes D$, let $u(B)$ be the composite:

$$C\otimes D \xrightarrow{u(C)\otimes u(D)} I\otimes I \xrightarrow{b} I.$$

QED.

For the purposes of our inductive proof we introduce for each object T a non-negative integer r(T), called its rank, defined by the following inductive rules:

$$r(I) = 0$$
$$r(\underline{1}) = 1$$
$$r(T\otimes S) = r(T) + r(S)$$
$$r([T,S]) = r(T) + r(S) + 1$$
$$r(\phi T) = rT + 1$$

Note that $rT = 0$ if and only if T is a constant integral ϕ-free object. If $f: T \to S$, we say the rank of f, r(f), is $r(T) + r(S)$.

Lemma 4.5: *If $T \to S$ is central then $r(T) = r(S)$.*

Proof: Use Lemma 6.3 of [5].

Proposition 4.6: (The Cut-Elimination Theorem) *If the morphisms $h: T \to S$ and $k: S\otimes U \to V$ of N are constructible then so is the composite morphism*

$$T\otimes U \xrightarrow{h\otimes 1} S\otimes U \xrightarrow{k} V .$$

Moreover, the graphs Γk and $\Gamma(h\otimes 1)$ are compatible, and the $\mathcal{D}$-graphs Δk and $\Delta(h\otimes 1)$ are compatible.

Proof: The graphs and $\mathcal{D}$-graphs are shown to be compatible by considering the same arguments as used in each case below. Thus we shall omit any further reference to the graphs or $\mathcal{D}$-graphs in the proof.

We use the same double induction as in [5]. Prime objects are those of the form $\underline{1}$, $[T,S]$ and S. We need only consider the following cases:

Case 7 h is of type ϕ, k is of type $\otimes$. This is considered in the same way as Case 5.

Case 8 h is of type π, k is of type ϕ. Since there is no central morphism: $[A,B]\otimes U \to K'_n(\phi C_1, \ldots, \phi C_n)$ this case does not exist.

Case 9 h is of type ϕ, k is of type $\langle\ \rangle$. We have

$$T\otimes U \xrightarrow{h\otimes 1} \phi A\otimes U \xrightarrow{x} ([B,C]\otimes D)\otimes E \xrightarrow{g(\langle f\rangle\otimes 1)} V.$$

There are two subcases:

Subcase 1: ϕA is associated via x with a prime factor of D. This case is considered in the same way as Case 6 Subcase 2.

Subcase 2: ϕA is associated via x with a prime factor of E. This case is considered in the same way as Case 6 Subcase 3.

Case 10: h and k are both of type ϕ. We have $k(h\otimes 1) =$

$$\begin{array}{c}
K'_m(K'_n(\phi A_1, \ldots, \phi A_n), \phi B_2, \ldots, \phi B_m) \\
\downarrow K'_m(\bar{\phi}_n, 1, \ldots, 1) \\
K'_m(\phi(K_n(A_1, \ldots, A_n)), \phi B_2, \ldots, \phi B_m) \\
\downarrow K'_m(\phi f, 1, \ldots, 1) \\
K'_m(\phi B_1, \phi B_2, \ldots, \phi B_m) \\
\downarrow \bar{\phi}_m \\
\phi(K_m(B_1, \ldots, B_m)) \\
\downarrow \phi g \\
\phi C
\end{array}$$

where $m > 1$. But $\bar{\phi}_m . K'_m(\phi f, 1, \ldots, 1) = \phi(K_m(f, 1, \ldots, 1)).\bar{\phi}_m$, trivially if $m = 1$, and by naturality of $\tilde{\phi}$ if $m > 1$. Also

$g.K_m(f, 1, \ldots, 1)$ is constructible by induction, and $\bar{\phi}_m.K'_m(\bar{\phi}_n, 1, \ldots, 1) = \bar{\phi}_{m+n-1}$ by §3. Thus $k(h\otimes 1) = \phi(g.K_m(f, 1, \ldots, 1))\bar{\phi}_{m+n-1}$, so is of type ϕ. QED.

Theorem 4.7: All morphisms of N are constructible.

Proof: As the proof of Theorem 6.5 of [5] except we need to show ϕf, $\tilde{\phi}$ and ϕ^{o} are constructible. In CB6 we obtain:

(1) ϕf, by letting $n = 1$
(2) $\tilde{\phi}$, by letting $n = 2$ and $f = 1$
(3) ϕ^{o}, by letting $n = 0$ and $f = 1$.

QED.

Corollary 4.8: Γ and Δ take their values in G and $\mathcal{D}$ respectively. In particular, N is a club.

Proof: Immediate by induction on rank, in view of Proposition 4.4.

The prime objects of type A are $\underline{1}$ and $[T,S]$. The prime objects of type B are $\underline{1}'$, $[T,S]'$ and ϕR. Every object T of type A (respectively B) may be written as $S(R_1, \ldots, R_n)$ where each R_i is a prime object of type A (respectively B), and S is an integral ϕ-free object of type A (respectively B). An integral ϕ-free object S is said to be reduced if S is either I (respectively I') or just formed by $\underline{1}$ and $\otimes$ (respectively $\underline{1}'$ and $\otimes'$). $T = S(R_1, \ldots, R_n)$ is said to be reduced if S is reduced, and if P_i is reduced whenever $R_i = \phi P_i$.

Lemma 4.9: Given any object T, we can find a reduced object T' and a central isomorphism $z: T \to T'$ in N.

Proof: Use the result of Lemma 7.1 of [5].

Lemma 4.10: In proposition 4.4 we can suppose that the objects

A,B,C,D in (2); A and D in (4), and A_1, B in (5) are reduced.

Proof: As Lemma 7.2 of [5].

We now define proper objects. I, I', $\underline{1}$ and $\underline{1}$' are proper. If T and S are proper, then T⊗S and T⊗'S are proper. If T and S are proper, then [T,S] and [T,S]' are proper, unless S is constant and T is not constant, or unless S is ϕ-free and T is not ϕ-free. If S is proper, then ϕS is proper. Observe that every constant ϕ-free object is proper; that if [T,S] is proper then T and S are proper; that T⊗S is proper if and only if T and S are proper; and that if ϕS is proper then S is proper - whence T is proper if and only if each of its prime factors is proper.

Remark: Note that again for convenience we have been omitting the primes on ⊗', [,]', etc. Thus ⊗ stands for ⊗ and/or ⊗'.

Lemma 4.11: Suppose h: T → S is central in N. Then if either T or S is proper so is the other.

Proof: Use Lemma 7.3 of [5].

Lemma 4.12: Let h: T → S in N. If T is proper and:

(1) S is constant, then T is constant;

(2) S is ϕ-free, then T is ϕ-free.

Proof: The proof of (1) is the same as that of Proposition 7.4 of [5]. Except we need consider $h = \phi f.\bar{\phi}_n$: $K_n'(\phi A_1, \ldots, \phi A_n) \to \phi B$. But ϕB is constant because S is. Since T is proper, each A_i is proper. Thus $\phi(K_n(A_1, \ldots, A_n))$ is proper. By induction $\phi(K_n(A_1, \ldots, A_n))$ is constant. Thus T is constant.

(2) is proved in the same way. QED.

We now show how to eliminate constant ϕ-free prime factors from an object T.

Lemma 4.13: Given an object T, we can find an object S with $rS \leq rT$, and an isomorphism $f: T \to S$ in N such that:

(1) S is reduced

(2) S has no constant ϕ-free prime factors, its prime factors being precisely those of T which are non-constant or not ϕ-free.

(3) If T is proper, so is S.

(4) There is a constant ϕ-free object R, and a central isomorphism $x: T \to S \otimes R$, with $\Gamma x = \Gamma f$ and $\Delta x = \Delta f$.

Proof: As Lemma 7.5 of [5].

Proposition 4.14: Let $h: P \otimes Q \to M \otimes N$ be a morphism of N, where P,Q,M,N are proper. Suppose $\Gamma h = \xi \otimes \eta$ and $\Delta h = \xi' \otimes \eta'$ for $\xi \in G(\Gamma P, \Gamma M)$, $\eta \in G(\Gamma Q, \Gamma N)$, $\xi' \in D(\Delta P, \Delta M)$ and $\eta' \in D(\Delta Q, \Delta N)$. Then there exist $p: P \to M$ and $q: Q \to N$ in N such that $h = p \otimes q$, $\Gamma p = \xi$, $\Gamma q = \eta$, $\Delta p = \xi'$ and $\Delta q = \eta'$.

Proof: We use induction on r(h). By Lemma 4.13 we may suppose each of P,Q,M,N to be reduced and have prime factors which are not both constant and ϕ-free.

Suppose h is central. If h is of type A, then ξ' and η' equal 1, and p and q, satisfying the proposition, exist by Proposition 7.6 of [5]. Suppose h is of type B. Suppose the prime factorisation of $P \otimes Q$ is $R(S_1, \ldots, S_n)$. Then h may be written as:

$$P \otimes Q = R(S_1, \ldots, S_n) \xrightarrow{f = 1(f_1, \ldots, f_n)} R(S_1', \ldots, S_n') = M' \otimes N' \xrightarrow{g_1 \otimes g_2} M \otimes N$$

where g_1, g_2 are generated by a', a'^{-1}, b', b'^{-1} and c'; and $f_i: S_i \to S_i'$ is ϕk_i if $S_i = \phi T_i$ where k_i is central of type A, and f_i is 1 otherwise. But f may be written as $f' \otimes f'': P \otimes Q \to M' \otimes N'$. Let $p = g_1 f'$ and $q = g_2 f''$.

Suppose h is of type $\otimes$. Form X,Y,U,V,X',Y', U' and V' as in Proposition 7.6 of [5]. Define a graph ρ: $\Gamma X \to \Gamma X'$ as the restriction of Γh. Define a $\mathcal{D}$-graph ρ': $\Delta X \to \Delta X'$ as the restriction of Δh. Similarly define graphs σ: $\Gamma Y \to \Gamma Y'$, τ: $\Gamma U \to \Gamma U'$, κ: $\Gamma V \to \Gamma V'$, and $\mathcal{D}$-graphs σ': $\Delta Y \to \Delta Y'$, τ': $\Delta U \to \Delta U'$, κ': $\Delta V \to \Delta V'$. The rest of this case is as in Proposition 7.6 of [5].

If h is of type π or of type < >, we use the same method as in Proposition 7.6 of [5] except we need to consider graphs and $\mathcal{D}$-graphs as above.

If h is of type ϕ, let h be the composite:

$$P\otimes Q \xrightarrow{x} K_n'(\phi A_1, \ldots, \phi A_n) \xrightarrow{\overline{\Phi}_n} \phi(K_n(A_1, \ldots, A_n)) \xrightarrow{\phi f} \phi B \xrightarrow{y} M\otimes N.$$

We may suppose M = ϕB and N = I. Each prime factor of P$\otimes$Q is of the form ϕT, so $\Delta(P\otimes Q) = \{+, +, \ldots, +\}$. Suppose Q $\neq$ I. Then ΔQ^+ is non--empty, and ΔQ^- is empty. By the form above of h, Δh maps each element of ΔQ^+ to $\Delta Q^- \cup \Delta N^+ = \emptyset$. We thus have a contradiction so Q = I. We can therefore let q be 1 and p be the composite:

$$P \xrightarrow{b^{-1}} P\otimes I \xrightarrow{h} M\otimes I \xrightarrow{b} M.$$

QED.

Proposition 4.15: Let f: A$\otimes$B $\to$ C be a morphism in N, where A,B,C are proper objects. Suppose for each x ε ΓB^+, y ε ΔB^+, that Γx ε ΓB^- and Δy ε ΔB^-. Then B is constant and ϕ-free.

Proof: Use Propositions 4.12 and 4.14 in the proof of Proposition 7.7 of [5].

Proposition 4.16: Let h: ([Q,M]$\otimes$P)$\otimes$ N $\to$ S be a morphism between proper objects in N, with [Q,M] not a constant ϕ-free object. Suppose that Γh is of the form $\eta(\langle\xi\rangle \otimes 1)$ for graphs ξ: $\Gamma P \to \Gamma Q$,

$\eta\colon \Gamma(M\otimes N) \to \Gamma S$. Suppose that Δh is of the form $\eta'(\langle\xi'\rangle \otimes 1)$ for $\mathcal{D}$-graphs $\xi'\colon \Delta P \to \Delta Q$, $\eta'\colon \Delta(M\otimes N) \to \Delta S$. Suppose that there do not exist objects F,G,E,H and a central morphism $x\colon P \to ([F,G]\otimes E)\otimes H$ such that ξ,ξ' can be written in the forms

$$\Gamma P \xrightarrow{\Gamma x} \Gamma(([F,G]\otimes E)\otimes H) \xrightarrow{\rho(\langle\sigma\rangle \otimes 1)} \Gamma Q,$$

$$\Delta P \xrightarrow{\Delta x} \Delta(([F,G]\otimes E)\otimes H) \xrightarrow{\rho'(\langle\sigma'\rangle \otimes 1)} \Delta Q$$

respectively for graphs ρ, σ and $\mathcal{D}$-graphs ρ', σ'. Then there exist morphisms $p\colon P \to Q$, $q\colon M\otimes N \to S$ in N, such that $h = q(\langle p\rangle \otimes 1)$, where $\Gamma p = \xi$, $\Gamma q = \eta$, $\Delta p = \xi'$, $\Delta q = \eta'$.

Proof: We proceed as in Proposition 7.8 of [5] except that we need consider both graphs and $\mathcal{D}$-graphs, and we need the case where h is of type ϕ. But h cannot be of type ϕ since $[Q,M]$ cannot be associated with any prime factor in the domain of a morphism of type ϕ. QED

Proposition 4.17: Let $h\colon K_n'(\phi A_1, \ldots, \phi A_n) \to \phi B$ be a morphism between proper objects in N. Then h may be written:

$$K_n'(\phi A_1, \ldots, \phi A_n) \xrightarrow{x} K_n'(\phi C_1, \ldots, \phi C_n) \xrightarrow{\overline{\phi}_n} \phi(K_n(C_1, \ldots, C_n)) \xrightarrow{\phi f} \phi B$$

for central x.

Proof: Obviously h cannot be of types π or $\langle\ \rangle$. If h is of type ϕ, h is certainly in the desired form.

Suppose h is central. Then $n = 1$. Thus $h = \phi f$ where f is central. So $h = \phi f.\ \overline{\phi}_1.1$ is in the desired form.

Suppose h is of type $\otimes$, so h may be written:

$$K_n'(\phi A_1, \ldots, \phi A_n) \xrightarrow{y} C\otimes D \xrightarrow{g\otimes k} E\otimes F \xrightarrow{z} \phi B.$$

We may suppose E to be ϕB, and F to be I. Suppose for some i, ϕA_i is associated via y with a prime factor of D. Thus ΔD^+ is not empty. By the form $z(g\otimes k)y$ of h, ΔD^+ is mapped to $\Delta D^- \cup \Delta F^+ = \emptyset$. This a contradiction so no factor ϕA_1 is associated with D, thus D = I and k is trivial, so h cannot be of type $\otimes$. QED.

Theorem 4.18: Let $h, h': T \to S$ be two morphisms between proper objects in N, such that $\Gamma h = \Gamma h'$ and $\Delta h = \Delta h'$. Then $h = h'$.

Proof: We use induction on $r(h) = r(h')$. By Lemma 4.13, we may suppose that none of the prime factors of T or of S is constant ϕ-free.

Suppose h and h' are both central. h may be written as

$$T = P(R_1, \ldots, R_n) \xrightarrow{f = 1(f_1,\ldots,f_n)} P(Q_1, \ldots, Q_n) \xrightarrow{g} S$$

where P is an integral ϕ-free object; $R_1, \ldots, R_n$ are the prime factors of T; g is generated by a, a^{-1}, b, b^{-1}, c if h is of type A, and by $a', a'^{-1}, b', b'^{-1}, c$ if h is of type B; and $f_i: R_i \to Q_i$ is 1 unless R_i is ϕM_i in which case f_i is ϕk_i for central k_i of type A. Since h' is of the same form, it is sufficient to show that $x, y: M \to N$ equal the identity, where $\Gamma x = \Gamma y$, $\Delta x = \Delta y$, and where x is of the same form as f, and y is of type B and generated by $a', a'^{-1}, b', b'^{-1}, c'$. But this is so by applying Theorem 3.1. We may now suppose that h is one of the other types.

If h is of type π or of type $\otimes$ we proceed as in Theorem 2.4 of [5].

Suppose h is of type ϕ. Then h is

$$T \xrightarrow{x} K'_n(\phi A_1, \ldots, \phi A_n) \xrightarrow{\bar{\phi}_n} \phi(K_n(A_1, \ldots, A_n)) \xrightarrow{\phi f} \phi B \xrightarrow{y} S$$

By Proposition 4.17, $y^{-1}h'x^{-1} = y^{-1}hx^{-1}: K'_n(\phi A_1, \ldots, \phi A_n) \to \phi B$ may be written

$$K'_n(\phi A_1, \ldots, \phi A_n) \xrightarrow{z} K'_n(\phi C_1, \ldots, \phi C_n) \xrightarrow{\bar{\phi}_n} \phi(K_n(C_1,\ldots,C_n)) \xrightarrow{\phi g} \phi B$$

$$= K_n'(\phi A_1, \ldots, \phi A_n) \xrightarrow{\bar{\Phi}_n} \phi(K_n(A_1, \ldots, A_n)) \xrightarrow{\phi(gw)} \phi B,$$

for a central w, by Theorem 3.1. Since $\Gamma h = \Gamma h'$, $\Gamma(y^{-1}h'x^{-1}) = \Gamma(y^{-1}hx^{-1})$, i.e. $\Gamma(\phi f.\bar{\Phi}_n) = \Gamma(\phi(gw).\bar{\Phi}_n)$. Thus $\Gamma(\phi f) = \Gamma(\phi(gw))$ since $\Gamma(\bar{\Phi}_n) = 1$. Thus $\Gamma f = \Gamma(gw)$. Since f and gw are type A morphisms, $\Delta f = \Delta(gw)$. By induction f = gw, hence h = h'.

Suppose h is of type < >, say h is

$$T \xrightarrow{x} ([B,C]\otimes A)\otimes D \xrightarrow{\langle f\rangle \otimes 1} C\otimes D \xrightarrow{g} S.$$

It may happen that Γf and Δf are of the forms:

$$\Gamma A \xrightarrow{\Gamma y} \Gamma(([F,G]\otimes E)\otimes H) \xrightarrow{\rho(\langle\sigma\rangle \otimes 1)} \Gamma B, \quad \text{and}$$

$$\Delta A \xrightarrow{\Delta y} \Delta(([F,G]\otimes E)\otimes H) \xrightarrow{\rho'(\langle\sigma'\rangle\otimes 1)} \Delta B,$$

for a central morphism $y: A \to ([F,G]\otimes E)\otimes H)$, graphs ρ,σ, $\mathcal{D}$-graphs ρ',σ'. Using the naturality of < > (which follows from the naturality of e), it follows that $\Gamma h = \tau(\langle\sigma\rangle\otimes 1)\psi$, $\Delta h = \tau'(\langle\sigma'\rangle\otimes 1)\psi'$ for graphs τ,ψ, $\mathcal{D}$-graphs τ',ψ'. But $\psi = \Gamma z$ and $\psi' = \Delta z$ where z is the central morphism:

$$T \xrightarrow{x} ([B,C]\otimes A)\otimes D \xrightarrow{1\otimes y} [B,C]\otimes(([F,G]\otimes E)\otimes H)$$

$$\xrightarrow{ac} ([F,G]\otimes E)\otimes(H\otimes[B,C]).$$

Possibly σ and σ' are of the forms:

$$\Gamma E \xrightarrow{\Gamma u} \Gamma(([X,Y]\otimes Z)\otimes W \xrightarrow{\kappa(\langle\lambda\rangle\otimes 1)} \Gamma F,$$

$$\Delta E \xrightarrow{\Delta u} \Delta(([X,Y]\otimes Z)\otimes W) \xrightarrow{\kappa'(\langle\lambda'\rangle\otimes 1)} \Delta F$$

for a central morphism $u: E \to ([X,Y]\otimes Z)\otimes W$, graphs κ,λ, and $\mathcal{D}$-graphs κ',λ'. But E has strictly fewer prime factors than A, since [F,G] is a prime factor of A but not of E; Z has strictly fewer prime factors than E; and so on. Thus this process terminates, and ultimately we have expressions for Γh and Δh of the form:

$$\Gamma T \xrightarrow{\Gamma v} \Gamma(([Q,M]\otimes P)\otimes N) \xrightarrow{\eta(\langle\xi\rangle\otimes 1)} \Gamma S,$$

$$\Delta T \xrightarrow{\Delta v} \Delta(([Q,M]\otimes P)\otimes N) \xrightarrow{\eta'(\langle\xi'\rangle\otimes 1)} \Delta S$$

where v is central, and ξ, ξ' are as in the Statement of Proposition 4.16. Moreover $[Q,M]$ is not constant ϕ-free since T has no constant ϕ-free prime factors. From Proposition 4.16 applied to hv^{-1} and $h'v^{-1}$ we conclude that $hv^{-1} = q(\langle p\rangle\otimes 1)$ and $h'v^{-1} = q'(\langle p'\rangle\otimes 1)$ for $p,p' : P \to Q$ and $q,q' : M\otimes N \to S$, with $\Gamma p = \Gamma p'$, $\Gamma q = \Gamma q'$, $\Delta p = \Delta p'$, $\Delta q = \Delta q'$. It follows from the inductive hypothesis that $p = p'$ and $q = q'$, so that $h = h'$.

This completes the proof of Theorem 4.18.

REFERENCES

[1] S. Eilenberg and G.M. Kelly, Closed categories, in: Proc. Conf. on Categorical Algebra, La Jolla, 1965 (Springer-Verlag, 1966) pp 421-562.

[2] D.B.A. Epstein, Functors between tensored categories, Invent. Math. 1 (1966) 221-228.

[3] G.M. Kelly, Many-variable functorial calculus. I. (in this volume).

[4] G.M. Kelly, An abstract approach to coherence. (in this volume).

[5] G.M. Kelly and S. Mac Lane, Coherence in closed categories, Journal of Pure and Applied Algebra 1 (1971) 97-140.

[6] S. Mac Lane, Natural associativity and commutativity, Rice University Studies 49 (1963) 28-46.

A CUT-ELIMINATION THEOREM

G.M. Kelly

University of New South Wales, Kensington 2033, Australia.

Received May 22, 1972

1. Introduction

We assume familiarity with the earlier paper [2] in this volume, and now refer particularly to §4.2 and §4.3 of that paper. Let $(\mathcal{B},\rho,\mathcal{D},\sigma)$ be as there the theory of an extra structure on a category or on a Λ-indexed polycategory, in the mixed-variance case with natural transformations of the every-variable-twice kind; and let K be the free model on Λ as constructed there.

One problem, as we saw, is that composable morphisms of K may have incompatible graphs; then K is not a club, and the free model on A cannot be written as $K \circ A$. Even when K <u>is</u> a club, it may not be easy to prove this fact directly from the construction of K given there.

Another problem comes from the fact that, if $R \to S \to T$ is a composite in K, the object S may have more variables (that is, higher type) than either R or T; this forms a barrier to certain kinds of inductive argument, which, like the incompatibility problem, is absent in the purely covariant case.

Both of these problems were overcome for particular theories in Kelly - Mac Lane ([3] and [4]) and in Lewis [6] by <u>cut-elimination</u> results, inspired by that of Lambek [5] and hence ultimately by the work of Gentzen [1]. Such a result asserts that the morphisms of K can be built up from simpler ones by a set of processes not involving composition (the analogue of Gentzen's "cut"). In favourable cases this restores the possibility of proofs by induction (on some measure of complexity of the objects T of K); it may allow in particular an inductive proof that incompatibility does not occur, as well as an inductive proof of some coherence assertion about K.

The purpose of this paper is to prove a general cut--elimination theorem, with the corollary that K is a club, when the theory $(\mathcal{B},\rho,\mathcal{D},\sigma)$ is of a certain special kind. Namely, we suppose that we begin with a purely covariant theory, which corresponds therefore to some club G in $\underline{\underline{Cat}}/\underline{\underline{Q}}$ where $\underline{\underline{Q}} = \underline{\underline{P}} \circ \Lambda \times \Lambda$; that we choose a family $\{P^\alpha\}$ of objects of G, where P^α has type $(n^\alpha[\lambda_1^\alpha, \ldots, \lambda_{n^\alpha}^\alpha], \mu^\alpha)$ with $n^\alpha \geq 1$; and that we choose for each α an $i^\alpha \in n^\alpha$. Then a model of our theory is to be a model A of G such that, if

$$|P^\alpha|: A_{\lambda_1} \times \ldots \times A_{\lambda_{n^\alpha}^\alpha} \to A_{\mu^\alpha}$$

is the functor corresponding to P^α, then the partial functor

$$|P^\alpha|(A_1, A_2, \ldots, A_{i^\alpha-1}, -, A_{i^\alpha+1}, \ldots, A_{n^\alpha}): A_{\lambda_{i^\alpha}^\alpha} \to A_{\mu^\alpha}$$

admits a right adjoint.

For simplicity of exposition and notation, we shall prove the theorem in the case when the family $\{P^\alpha\}$ is replaced by a single object P. The reader will see at once that the proof carries over to the more general case; we avoid it only because the notational difficulties become horrific. The reader should further observe that we called $\{P^\alpha\}$ a family advisedly; there is no reason for the P^α to be distinct; if G corresponds to the theory of monoidal categories, and if we take $P^1 = P^2 = \otimes$ while $i^1 = 1$ and $i^2 = 2$, then a model of our theory is a biclosed category - that is, a monoidal category in which both $A\otimes -$ and $-\otimes B$ have right adjoints. We might as well suppose that the pairs (P^{α}, i^{α}) are distinct - but it doesn't really make any difference; if we want to posit two right adjoints for $A\otimes -$, with different names, then they'll be isomorphic, but that won't affect our result.

We further simplify by taking A to consist of one element - the single-category case; again it will be clear from the proof that this makes no difference. Finally we simplify the notation by supposing that the "active" variable in $|P|$ is the first one; we take P to have type $1 + k$ where $k \geq 0$, so that $|P|: A^{1+k} \to A$, and then a model of our theory is a model A of G such that

$$|P|(-, B_1, \ldots, B_k): A \to A$$

has a right adjoint

$$|Q|(B_1, \ldots, B_k, -): A \to A.$$

Here $|Q|$ is a functor $A^{op} \times \ldots \times A^{op} \times A \to A$, contravariant in every variable except the last. (Think of P as "$\otimes$" and Q as "internal hom".)

We spoke loosely in describing a model of our theory as a model of G in which $|P|$ admits a right adjoint $|Q|$; if we want our theory to be of the desired equational form $(\mathcal{B},\rho,\mathcal{D},\sigma)$ we must make Q, of type $(k + 1)[-- \ldots - +]$, part of the data; we must also make the unit and counit of the adjunction,

$$d: A \to Q(B_1, \ldots, B_k, P(A, B_1, \ldots, B_k)),$$

$$e: P(Q(B_1, \ldots, B_k, A), B_1, \ldots, B_k) \to A,$$

part of the data; and we must make the triangular axioms on d and e, asserting that they constitute an adjunction, part of the axioms.

2. Formal description of the theory

$\mathcal{B}$, then, consists of $\underline{\underline{Ob}}\ G$ together with one extra element Q of the appropriate type. The free discrete club T generated by $\mathcal{B}$ has an object $\underline{1}$; has objects $T(Y_1, \ldots, Y_n)$ for $T \in G$, $Y_i \in T$; and has objects $Q(Y_1, \ldots, Y_{k+1})$ for $Y_i \in T$. We identify $T \in G$ with $T(\underline{1}, \ldots, \underline{1})$ and Q with $Q(\underline{1}, \ldots, \underline{1})$. The discrete club of objects of K is obtained from T by imposing the following relations ρ: the "substitution" $T(S_1, \ldots, S_n)$ in $\underline{\underline{Ob}}\ K$, when $T, S_i \in G$, is equal to the $T(S_1, \ldots, S_n)$ of G; and the $\underline{1}$ of K is equal to that of G.

Call an object of K prime if it is either $\underline{1}$ or else of the form $Q(Y_1, \ldots, Y_{k+1})$. Then it is clear from the above that every object K of K has a unique expression as

$$K = T(X_1, \ldots, X_n) \tag{2.1}$$

where $T \in G$ and the X_i are prime. We call (2.1) the prime factorization of K. (The prime factorization of $T \in G$ is $T(\underline{1}, \ldots, \underline{1})$; in particular that of $\underline{1}$ is $\underline{1}(\underline{1})$; and the prime factorization of $Z = Q(Y_1, \ldots, Y_{k+1})$ is $\underline{1}(Z)$.) We call $X_1, \ldots, X_n$ the prime factors of K.

$\mathcal{D}$ consists of $\underline{\underline{Mor}}\ G$ with two extra elements

$$d: \underline{1} \to Q(\underline{1}, \ldots, \underline{1}, P),$$

$$e: P(Q, \underline{1}, \ldots, \underline{1}) \to \underline{1},$$

of appropriate graphs. Finally the relations σ assert that the composite fg in K when $f, g \in G$ is their composite in G; that 1_T in K is 1_T in G if $T \in G$; that substitution in K of morphisms in G agrees with that in G; and that d and e satisfy the triangular equations for an adjunction.

Since K is a fortiori a G-category, there is a unique map $\phi: G \to K$ of G-categories sending $\underline{1}$ to $\underline{1}$. I conjecture that ϕ is necessarily faithful; in the absence of a proof of this, however, we shall now for simplicity replace G by its image H under ϕ. The kernel-congruence π of ϕ is clearly a club-congruence, so $H = G/\pi$ is again a covariant club with the same objects as G; we identify H with a subcategory of K.

The type $n[+, +, \ldots, +]$ of an object of K that lies in H will usually be abbreviated to n. An object of K of type $0[\;]$, or an object of H of type 0, is said to be constant. The prime factorization of a constant T of H is $T(\;)$.

3. Central morphisms

Define a central morphism of K to be one of the form $x(X_1, \ldots, X_n): T(X_{\xi 1}, \ldots, X_{\xi n}) \to S(X_1, \ldots, X_n)$ where $T, S \in H$, where the X_i are prime, and where $x: T \to S$ in H with graph $\Gamma x = \xi$. It is clear that the composite of central morphisms is central, and that every identity morphism is central; so the centrals form a subcategory of K.

Lemma 3.1 *If $x: T \to S$ in H with graph ξ, then $x(Y_1, \ldots, Y_n): T(Y_{\xi 1}, \ldots, Y_{\xi n}) \to S(Y_1, \ldots, Y_n)$ is central whether the Y_i are prime or not.*

Proof Let the prime factorizations of the Y_i be $Y_1 = R_1(X_1, \ldots, X_{m_1}), \ldots, Y_n = R_n(\ldots X_m)$. Then $x(Y_1, \ldots, Y_n) = x(R_1, \ldots, R_n)(X_1, \ldots, X_m)$, and $x(R_1, \ldots, R_n)$ is in H.

Lemma 3.2 *If $z_i: Y_i \to Z_i$ is central for each i, and if $R \in H$, then $R(z_1, \ldots, z_n): R(Y_1, \ldots, Y_n) \to R(Z_1, \ldots, Z_n)$ is central.*

<u>Proof</u> Let $z_1 = x_1\ (X_1, \ldots, X_{m_1})\colon T_1(X_{\xi_1 1}, \ldots) \to S_1(X_1, \ldots)$ etc., where $x_i\colon T_i \to S_i$ is in H and the X_i are prime. Then $R(z_1, \ldots, z_n) = R(x_1, \ldots, x_n)(X_1, \ldots, X_m)$, and $R(x_1, \ldots, x_n)$ is in H.

Observe that if $z\colon Y \to Z$ is central then Y and Z have the same prime factors, to within order.

<u>4. Statement of the theorem</u>

For $Y, W_1, Z \in K$ write π for the adjunction:

(4.1) $\pi\colon K(P(Y, W_1, \ldots, W_k), Z) \cong K(Y, Q(W_1, \ldots, W_k, Z))$.

Thus $\pi(f)$ is in fact the composite

(4.2) $$Y \xrightarrow{d} Q(W_1, \ldots, W_k, P(Y, W_1, \ldots, W_k)) \xrightarrow{Q(1,\ldots,1,f)} Q(W_1, \ldots, W_k, Z)$$

where d stands for $d(Y, W_1, \ldots, W_k)$.

For $Z \in K$ and for $f_i\colon Y_i \to W_i$ in K $(1 \le i \le k)$, write $\langle Z; f_1, \ldots, f_k\rangle$, or just $\langle f_1, \ldots, f_k\rangle$ for short, for the composite

(4.3) $$P(Q(W_1,\ldots,W_k,Z),Y_1,\ldots,Y_k) \to P(Q(W_1,\ldots,W_k,Z),W_1,\ldots,W_k) \xrightarrow{e} Z$$

where the first morphism is $P(1, f_1, \ldots, f_k)$ and where e stands for $e(W_1, \ldots, W_k, Z)$. As a still shorter notation for (4.3) we shall also use

(4.4) $$\langle f\rangle\colon P(Q(W, Z), Y) \to Z.$$

Define the set of <u>constructible</u> morphisms of K to be the smallest set satisfying C1–C4 below:

C1. Every central morphism is constructible.

C2. Let $T \in H$ with type $n \ge 2$, and let $f_i\colon Y_i \to Z_i$ be constructible for $1 \le i \le n$. Suppose for each i that Y_i and Z_i are not <u>both</u> constants of H (they are allowed to be constants of K!). Let a and b be central. Then the following composite is constructible:

$$K \xrightarrow[a]{} T(Y_1,\ldots, Y_n) \xrightarrow[T(f_1,\ \ldots,\ f_n)]{} T(Z_1,\ \ldots,\ Z_n) \xrightarrow[b]{} L.$$

C3. Let $T \in H$ with type 1, let $f\colon P(Y, W_1, \ldots, W_k) \to Z$ be constructible, and let a and b be central. Then the following composite is constructible:

$$K \xrightarrow[a]{} T(Y) \xrightarrow[T(\pi(f))]{} T(Q(W_1,\ \ldots,\ W_k,\ Z)) \xrightarrow[b]{} L.$$

C4. Let $T \in H$ with type $n\geq 1$, and let $u \in n$. Let $g\colon T(V_1,\ \ldots,\ V_n) \to L$ be constructible, and let $f_j\colon Z_j \to W_j$ be constructible for $1\leq j\leq k$. Let a be central. Then the following composite, where we have used the abbreviated notation (4.4) for (4.3), is constructible:

$$K \xrightarrow[a]{} T(V_1\ldots P(Q(W,V_u),Z)\ldots V_n) \xrightarrow[T(1\ldots\langle f\rangle\ldots 1)]{} T(V_1\ldots V_u\ldots V_n) \xrightarrow[g]{} L.$$

<u>Theorem 4.1</u> <u>Every morphism of K is constructible.</u>

We prove the theorem in §6 below. It is clearly inspired by §6 of [3]. The reader making a comparison should observe that ⊗ in [3] plays two roles which are separated here; namely the role of P, and also that of T in C2 and C4. Clearly if G corresponded to the theory (B',ρ',D',σ'), and if B' had no objects of type 1, it would be sufficient in C2 to suppose $T \in B'$; in [3] the only functor in B' of type $n\geq 2$ is ⊗, which is also our P. The reader should also notice that we treat differently objects T of H with type ≥ 2, and those of type 1; the first appear in C2, the second in C3; both can occur in C4. The objects of H of type 0 are of course covered by C1.

The morphisms of K constructed in C2, C3, C4 will be said to be respectively of types *fun*, *adj*, and *ev*, to suggest their methods of formation (apply a functor T; take an adjoint; apply evaluation).

In the case when there are many P^α with right adjoints Q^α, there are of course construction rule $C3^\alpha$ anc $C4^\alpha$ for each α.

5. K is a club

Define inductively a rank for the objects of K by setting

$$r(\underline{1}) = 1,$$

$$r(T(Y_1, \ldots, Y_n)) = r(Y_1) + \ldots + r(Y_n) \text{ for } T \in H,$$

$$r(Q(Y_1, \ldots, Y_{k+1})) = r(Y_1) + \ldots + r(Y_{k+1}) + 1;$$

in the first instance this is a rank on the objects of the free discrete club generated by $B = \underline{\underline{Ob}}H \cup \{Q\}$, but it is clearly compatible with the equivalence relation ρ and defines a rank on $\underline{\underline{Ob}}\, K$. If $K \to L$ is central, K and L have the same prime factors and hence the same rank. The only objects of rank 0 are the constants of H.

For a morphism $f: K \to L$ of K set $r(f) = r(K) + r(L)$. Then it is clear that:

Lemma 5.1 The morphism constructed in C2 has rank $> r(f_i)$ for each i; that constructed in C3 has rank $> r(f)$; and that constructed in C4 has rank $> r(f_j)$ for each j and also $> r(g)$.

The category K is augmented over $\underline{P}^*_o$, whose objects are types and whose morphisms are simple graphs. We recall that we compose incompatibles in $\underline{P}^*_o$ just by discarding any closed loops that occur. Our present purpose is better served, however, by introducing a new category $\underline{P}'$, whose objects are still the mixed-variance types, but whose morphisms $\tau \to \sigma$ consist of all the simple graphs together with one new element $*$. The composite of η and ξ in $\underline{P}'$ is their composite in $\underline{P}^*_o$ if η and ξ are compatible simple graphs; in all other cases it is $*$. That this is an associative composition, making $\underline{P}'$ a category, follows from the fact that, if ζ is compatible with η and $\zeta\eta$ with ξ, then η is compatible with ξ and ζ with $\eta\xi$; and conversely.

It is clear from the construction of K in §4.2 and §4.3 of [2] that it can be given an augmentation Γ' over $\underline{P}'$. Then to say that no composables are incompatible in K is to say that $\Gamma' f$ is never $*$ for $f \in K$.

The graph of a central morphism is a pure permutation (that is, any mates under it occur one in the domain and one in the codomain); it is therefore compatible with anything at all. The nature of the constructions C2-C4 shows that the constructed morphism does not have augmentation $*$ unless one of the hypothesis-morphisms does. Lemma 5.1 now gives an immediate proof, by induction on $r(f)$, that $\Gamma' f \neq *$ for $f \in K$; thus

Theorem 5.2 Composable morphisms of K have compatible graphs; K is a club.

6. Proof of the main theorem

Lemma 6.1 If h is constructible and u,v are central then vhu is constructible.

Proof Since constructibles are closed under composition, it is clear from the form of C1-C4 that hu is constructible. The same is clear for vh unless h is produced by C4; so the constructibility of vh clearly follows by induction on $r(h)$, in view of Lemma 5.1.

Lemma 6.2 If K, L are constants of H, any constructible $h: K \to L$ is central.

Proof Immediate from Lemma 5.1.

Lemma 6.3 Let $T \in H$ with type $n \geq 1$, and let $f_1: Y_1 \to Z_1$ be

<u>constructible for</u> $1 \le i \le n$. <u>Then</u> $T(f_1, \ldots, f_n)$ <u>is constructible</u>.

<u>Proof</u> Let those i for which Y_i and Z_i are both constants of H be m+1, m+2, ..., n say. Then

$$T(f_1,\ldots,f_n) = T(f_1,\ldots f_m,1,\ldots,1)\ T(1,\ldots,1,f_{m+1},\ldots,f_n).$$

The right-hand factor is central by Lemmas 6.2 and 3.2, and we can discard it by Lemma 6.1. The remaining factor $T(f_1, \ldots, f_m, 1, \ldots, 1)$: $T(Y_1, \ldots, Y_m, Z_{m+1}, \ldots, Z_n) \to T(Z_1, \ldots, Z_m, Z_{m+1}, \ldots, Z_n)$ is of the form $S(f_1, \ldots, f_m)$: $S(Y_1, \ldots, Y_m) \to S(Z_1, \ldots, Z_m)$ where $S = T(\underline{1}, \ldots, \underline{1}, Z_{m+1}, \ldots, Z_n) \in H$. If $m \ge 2$ this is constructible by C2; if m = 0 it is the identity and therefore central and constructible. It remains therefore to prove that, if $S \in H$ has type 1, S(h) is constructible wherever h is. Using Lemma 3.2 and the form of C1-C4, this admits an immediate proof by induction on r(h).

<u>Lemma 6.4</u> <u>Let</u> S <u>be an object of</u> H, <u>of type</u> m+1 <u>where</u> $m \ge 0$, <u>and let</u> $i \in m+1$. <u>Let</u> $h: K \to L$ <u>and</u> $t: S(M_1,\ldots,M_{i-1},L, M_i, \ldots, M_m) \to N$ <u>be constructible</u>. <u>Then the composite</u>

$$(6.1)\qquad S(M_1 \ldots K \ldots M_m) \xrightarrow[S(1\ \ldots\ h\ \ldots\ 1)]{} S(M_1 \ldots L \ldots M_m) \xrightarrow[t]{} N$$

<u>is constructible</u>.

We shall prove Lemma 6.4 in the next section; it is analogous to Proposition 6.4 of [3]. Assuming it we prove the main Theorem 4.1.

An instance of a morphism x of H is central by Lemma 3.1 and therefore constructible. An instance of d is of the form $\pi(1)$ and is therefore constructible by C3 (with $T = \underline{1}$). An instance of e is of the form <1, 1, ..., 1>, and is therefore constructible by C4 (with $T = \underline{1}$ and g = 1). If h is constructible its expansion T(1...h...1) for $T \in H$ is constructible by Lemma 6.3. If h and $f_1, \ldots, f_k$ are constructible so is $Q(f_1, \ldots, f_k, h)$, for this is the image under π of a composite h<f>, and the latter is constructible by C4 (with $T = \underline{1}$). The constructibles therefore contain all expanded instances of d, of e, and of the morphisms of H; since these generate K, and since the constructibles are closed under composition by the case m = 0 of Lemma 6.4, the constructibles constitute the whole of K.

7. Proof of Lemma 6.4

In the situation of Lemma 6.4, write

$$\sigma = r(K) + r(L) + r(N) + \Sigma\, r(M_i) \tag{7.1}$$

and write

$$\sigma_o = r(K) + r(L). \tag{7.2}$$

The proof is by a double induction; we suppose (6.1) to be constructible for all such situations with a lower σ, and also for all situations with the same σ but with a lower σ_o. We remind the reader that if $Y \to Z$ is central, then $r(Y) = r(Z)$, as we said in §5.

Since by Lemma 3.2 $S(1 \ldots h \ldots 1)$ is central when h is, the constructibility of (6.1) is immediate by Lemma 6.1 if either h or t is central. For non-central h and t, we consider cases according to their mode of construction by C2-C4. We simplify by observing that, if h is produced by C2, C3, or C4, we can ignore a, which merely puts a central factor $S(1 \ldots a \ldots 1)$ in front of (6.1); and that we can also ignore b (when h is produced by C2 or C3) by absorbing $S(1 \ldots b \ldots 1)$ into t. Similarly we can ignore factors b occuring at the end of t.

Case 1: h is of type *fun*

With h as in C2, but ignoring a and b, (6.1) becomes the composite

$$(7.3)\ R(M_1 \ldots Y_1 \ldots Y_n \ldots M_m) \xrightarrow[R(1\ldots f_1 \ldots f_n \ldots 1)]{} R(M_1 \ldots Z_1 \ldots Z_n \ldots M_m) \xrightarrow[t]{} N,$$

where $R \in H$ is $S(\underline{1}, \ldots, \underline{1}, T, \underline{1}, \ldots, \underline{1})$. Consider the composite

$$(7.4)\ R(M_1 \ldots Z_1 \ldots Z_{n-1}\ Y_n \ldots M_m) \xrightarrow[R(1\ldots f_n \ldots 1)]{} R(M_1 \ldots Z_1 \ldots Z_{n-1} Z_n \ldots M_m) \xrightarrow[t]{} N.$$

This is again of the form (6.1); but its σ has only a contribution $r(Z_1)$ where the σ of (6.1) itself, now written as (7.3), has a contribution $r(Y_1) + r(Z_1)$(we are using the fact that $n \geq 2$). So the σ of (7.4) is lower unless $r(Y_1) = 0$, that is, unless Y_1 is a constant of H. But in that case Z_1 is <u>not</u> a constant of H, and then the σ_o of (7.4), namely $r(Y_n)+r(Z_n)$, is less than that of (6.1) = (7.3), namely $r(Y_1) + \ldots + r(Y_n) + r(Z_1) + \ldots + r(Z_n)$. So by induction (7.4) is constructible.

Repetition of this argument if $n > 2$ yields the constructibility of $\bar{t}R(1\ldots f_{n-1} \ldots 1)$, where $\bar{t}$ is (7.4); and hence ultimately of

$$(7.5) R(M_1 \ldots Z_1\ Y_2 \ldots Y_n \ldots M_m) \xrightarrow[R(1\ldots f_2 \ldots f_n \ldots 1)]{} R(M_1 \ldots Z_1\ Z_2 \ldots Z_n \ldots M_m) \xrightarrow[t]{} N.$$

Writing t' for (7.5), we can write (7.3) as

$$(7.6)\ R(M_1 \ldots Y_1\ Y_2 \ldots Y_n \ldots M_m) \xrightarrow[R(1\ldots f_1 \ldots 1)]{} R(M_1 \ldots Z_1\ Y_2 \ldots Y_n \ldots M_m) \xrightarrow[t']{} N.$$

The σ of (7.6) is less than that of (6.1) = (7.3) unless $r(Z_n) = 0$; but in that case the σ_o of (7.6) is less than that of (6.1). So by induction (7.6) is constructible, as required.

<u>Case 2: h is of type *ev*</u>

With h as in C4 but ignoring a, we can write (6.1) as $t\ S(1 \ldots g \ldots 1)\ R(1 \ldots \langle f\rangle \ldots 1)$, where $R = S(\underline{1} \ldots T \ldots \underline{1}) \in H$.

The composite $t' = t\ S(1\ldots g\ldots 1)$ is constructible by induction, having a lower σ than (6.1) since $r(g) < r(h)$. Then (6.1) is $t'\ R(1\ldots\langle f\rangle\ldots 1)$ which is constructible by C4.

Case 3 : h is of type *adj*

We take h to be $G(\pi(s))\colon G(A) \to G(Q(B,C))$, where B stands for $B_1, \ldots, B_k$. We may as well take $G = \underline{\underline{1}}$, for we can replace S by $S(\underline{\underline{1}}\ \ldots\ G\ \ldots\ \underline{\underline{1}})$. Thus

(7.7) $$h = \pi(s)\colon A \to Q(B,C),$$

where

(7.8) $$s\colon P(A,B) \to C$$

is constructible. Next, we may as well suppose the M_i in (6.1) to be prime; for if their prime factorizations are $M_1 = R_1(X_1, \ldots)$ etc, we just replace S by $S(R_1\ \ldots\ \underline{\underline{1}}\ \ldots\ R_m)$. So the codomain of

(7.9) $$S(1\ldots h\ldots 1)\colon S(M_1\ldots A\ldots M_m) \to S(M_1\ldots Q(B,C)\ldots M_m)$$

is already expressed in its prime factorization.

Whether t is of type *fun*, *adj*, or *ev*, its first factor is a central a, which we shall write for the moment as

(7.10) $$a\colon S(M_1\ldots Q(B,C)\ldots M_m) \longrightarrow T(Y_1, \ldots, Y_n).$$

Let the prime factorizations of the Y_i be $Y_1 = R_1(X_1\ldots)$ etc., so that $T(Y_1\ldots Y_n) = T(R_1\ldots R_n)(X_1\ldots X_{m+1})$. Since a is central the primes X_i are $Q(B,C)$ and the M_i in some order, and a is $x(X_1\ldots X_{m+1})$ for some $x\colon S \to T(R_1\ldots R_n)$ in $\mathcal{H}$ with some graph ξ. Suppose for simplicity that ξ equates $Q(B,C)$ with the prime factor X_1 of Y_1, so that $Y_1 = R_1(Q(B,C), X_2, \ldots)$. Set $U = R_1(A, X_2, \ldots)$ and define $p\colon U \to Y_1$ by

(7.11) $$p = R_1(h, X_2, \ldots): U \to Y_1.$$

We have the commutative diagram

$$\begin{array}{ccc} S(M_1\ldots A\ldots M_m) & \xrightarrow{x(A,X_2\ldots X_{m+1})} & T(R_1\ldots R_n)(A,X_2\ldots X_{m+1}) \\ \downarrow{\scriptstyle S(1\ldots h\ldots 1)} & & \downarrow{\scriptstyle T(R_1\ldots R_n)(h,1,\ldots 1)} \\ S(M_1\ldots Q(B,C)\ldots M_m) & \xrightarrow[x(Q(B,C),X_2\ldots X_{m+1})]{} & T(R_1\ldots R_n)(Q(B,C),X_2\ldots X_{m+1}); \end{array}$$

that is,

(7.12) $$\begin{array}{ccc} S(M_1\ldots A\ldots M_m) & \xrightarrow{a'} & T(U, Y_2\ldots Y_n) \\ \downarrow{\scriptstyle S(1\ldots h\ldots 1)} & & \downarrow{\scriptstyle T(p,\ 1,\ \ldots 1)} \\ S(M_1\ldots Q(B,C)\ldots M_m) & \xrightarrow[a]{} & T(Y_1, Y_2\ldots Y_n), \end{array}$$

where we have written a' for the central $x(A, X_2\ldots X_{m+1})$.

More generally now, allowing for the fact that ξ may equate $Q(B,C)$ with any prime factor of any Y_1, we have expressed $aS(1\ \ldots\ h\ \ldots 1)$ by (7.12) as $T(1\ \ldots\ p\ \ldots\ 1)a'$, with a' central and with (in place of (7.11))

$$p: U \to Y_1$$

of the form

(7.13) $$R(X_1\ \ldots\ A\ \ldots\ X_q) \xrightarrow[R(1\ \ldots\ h\ \ldots\ 1)]{} R(X_1\ldots Q(B,C)\ldots X_q)$$

for some $R \in H$. Let us express the σ of (6.1) in these terms; since the X_j in (7.13) are some of the M_j; since the prime factors of the Y_u for $u \neq 1$ are the rest of the M_j; and since $K = A$ and $L = Q(B,C)$ — the σ of (6.1) is

(7.14) $$\sigma = r(A) + r(Q(B,C)) + \Sigma\, r(X_j) + \sum_{u\neq 1} r(Y_u) + r(N).$$

Since we can ignore a' by Lemma 6.1, we are reduced to proving the constructibility of $t\,T(1 \dots p \dots 1)$ where $t\colon T(Y_1 \dots Y_n) \to N$ is constructed by C2, C3, or C4 but <u>without</u> the a, which we have already allowed for, and <u>without</u> the b, which we can ignore. We proceed to the cases.

<u>Case 3(a)</u>: t <u>is of type *fun*</u>

Then $t = T(f_1 \dots f_n)$ and $t\,T(1 \dots p \dots 1)$ is $T(f_1 \dots f_i p \dots f_n)$. This is constructible by Lemma 6.3 provided $f_i p$ is constructible. But $f_i p$ is

$$(7.15)\quad R(X_1 \dots A \dots X_q) \xrightarrow[R(1\dots h\dots 1)]{} R(X_1 \dots Q(B,C) \dots X_q) \xrightarrow[f_i]{} Z_i;$$

this is of the form (6.1), and its σ is strictly less than (7.14); for $r(N) = \Sigma\, r(Z_u)$ and, n being ≥ 2, we have lost at least one contribution $r(Y_j) + r(Z_j)$ with $j \neq i$. Hence (7.15) is constructible by induction.

<u>Case 3(b)</u>: t <u>is of type *adj*</u>

Then $n = 1$, so $T(1 \dots p \dots 1)$ is just $T(p)$; and $t = T(\pi(f))$, so that $t\,T(1 \dots p \dots 1)$ is $T(\pi(f)p)$. By the naturality of π, the composite $\pi(f)p$ is $\pi(g)$ where g is the composite

$$(7.16)\qquad P(U,W) \xrightarrow[P(p,1)]{} P(Y,W) \xrightarrow[f]{} Z.$$

Setting $H = P(R,\underline{1} \dots \underline{1}) \in \mathcal{H}$ (recall that W stands for $W_1, \dots, W_k$), this equal by (7.13) to

$$(7.17)\quad H(X_1 \dots A \dots X_q, W) \xrightarrow[H(1\dots h\dots 1)]{} H(X_1 \dots Q(B,C) \dots X_q, W) \xrightarrow[f]{} Z.$$

But (7.17) is of the form (6.1), and its σ is less than (7.14) because $r(N) = r(Q(W,Z)) = r(W) + r(Z) + 1$. Hence $g = (7.17)$ is constructible by induction, whence $t\,T(1 \dots p \dots 1) = T(\pi(g))$ is constructible by C3.

<u>Case 3(c)</u> : t <u>is of type ev</u>

We take $t = g\,T(1 \dots \langle f\rangle \dots 1)$ as in C4, with a omitted. Here $Y_1, \dots, Y_n = V_1, \dots, P(Q(W, V_u), Z), \dots V_n$. We must distinguish three subcases:

<u>Subcase (i)</u>

Suppose the codomain Y_1 of p is one of the V_j for $j \neq u$; for simplicity suppose it is $Y_1 = V_1$. Then $t\,T(p,1, \dots, 1) =$ $g\,T(1 \dots\langle f\rangle\dots 1)\,T(p,1, \dots 1) = g\,T(p,1, \dots, 1)\,T(1\dots\langle f\rangle\dots 1)$. This will be constructible by C4 if $g\,T(p,1,\dots, 1)$ is. The latter is

$$(7.18) \qquad T(U,V_2, \dots, V_n) \xrightarrow[T(p,1,\dots,1)]{} T(V_1, V_2, \dots, V_n) \xrightarrow[g]{} N;$$

that is, setting $H = T(R, \underline{1}, \dots, \underline{1})$,

$$(7.19) \qquad H(X_1\dots A\dots X_q,V) \xrightarrow[H(1\dots h\dots 1)]{} H(X_1\dots Q(B,C)\dots X_q,V) \xrightarrow[g]{} N,$$

where $V = V_2, \dots, V_n$. This is of the form (6.1), and its σ is less than (7.14) because we have replaced the contribution $r(Y_u) = r(Q(W,V_u)) + r(Z)$ by $r(V_u)$, which is strictly less; hence it is constructible by induction.

<u>Subcase (ii)</u>

Suppose the codomain Y_1 of p is $P(Q(W,V_u),Z)$ and that the prime factor $Q(B,C)$ of Y_1 is $Q(W,V_u)$. Then U is $P(A,Z)$, and p is $P(h,1)\colon P(A,Z) \to P(Q(B,C),Z)$. It is easy to verify the commutativity of

$$\begin{array}{ccc}
T(V_1\ldots P(A,Z)\ldots V_n) & \xrightarrow{T(1\ldots P(1,f)\ldots 1)} & T(V_1\ldots P(A,B)\ldots V_n) \\
{\scriptstyle T(1\ldots P(h,1)\ldots 1)}\big\downarrow & & \big\downarrow{\scriptstyle T(1\ldots s\ldots 1)} \\
T(V_1\ldots P(Q(B,C),Z)\ldots V_n) & \xrightarrow[T(1\ldots <f>\ldots 1)]{} & T(V_1\ldots C\ldots V_n),
\end{array}$$

where $h = \pi(s)$ as in (7.7). Then $g\ T(1\ldots <f>\ldots 1)\ T(1\ldots p\ldots 1)$ is $g\ T(1\ldots s\ldots 1)\ T(1\ldots P(1,f)\ldots 1)$. First, $g\ T(1\ldots s\ldots 1)$ is of the form (6.1), and it is constructible by induction because its $\sigma = \sum_{j\neq u} r(V_j) + r(A) + r(B) + r(C) + r(N)$ is less than (7.14) since $r(Q(B,C)) = r(B) + r(C) + 1$. Next, writing g' for $g\ T(1\ldots s\ldots 1)$, and H for $T(\underline{\underline{1}}\ldots P\ldots\underline{\underline{1}})$, we have to prove the constructibility of

$$H(V_1\ldots A,Z_1\ldots Z_k\ldots V_n) \xrightarrow[H(1\ldots f_1\ldots f_k\ldots 1)]{} H(V_1\ldots A,B_1\ldots B_k\ldots V_n) \xrightarrow[g']{} N.$$

We prove this step-by-step as in Case 1; only we need not, as there, suppose $k \geq 2$, or use the side induction on σ_o; for

$$\sum_{j\neq u} r(V_j) + r(A) + r(Z) + r(B) + r(N)$$

is already less than (7.14), $r(Z)$ being the $\Sigma\ r(X_j)$ of (7.14).

Subcase (iii)

Suppose the codomain Y_1 of p is $P(Q(W,V_u), Z)$ and that the prime factor $Q(B,C)$ of Y_1 is a prime factor of some Z_j, say Z_1. Write Z_1' for Z_1 with the prime factor $Q(B,C)$ replaced by A. Then p is $P(1,r,1\ldots 1)$, where $r\colon Z_1' \to Z_1$ is of the form

$$R(1\ldots h\ldots 1)\colon R(X_1'\ldots A\ldots) \to R(X_1'\ldots Q(B,C)\ldots).$$

In this case (7.14) becomes

(7.20)
$$\sigma = r(A) + r(Q(B,C)) + r(Q(W,V_u)) + \Sigma\ r(X_j') + \sum_{j\neq 1} r(Z_j) + \sum_{j\neq u} r(V_j) + r(N).$$

It is easy to verify the commutativity of

$$\begin{array}{ccc}
T(V_1\ldots P(Q(W,V_u),Z_1',Z_2\ldots)\ldots V_n) & & \\
\downarrow{\scriptstyle T(1\ldots P(1,r,\ldots 1)\ldots 1)} & \searrow{\scriptstyle T(1\ \ldots\ \langle\bar{f}\rangle\ \ldots\ 1)} & \\
T(V_1\ldots P(Q(W,V_u),Z_1,Z_2\ldots)\ldots V_n) & \xrightarrow[T(1\ldots\langle f\rangle\ldots 1)]{} & T(V_1,\ldots,V_n)
\end{array}$$

where $\bar{f} = f_1r,\ f_2,\ \ldots,\ f_n$. Thus $t\ T(1\ldots p\ldots 1) = g\ T(1\ldots\langle\bar{f}\rangle\ldots 1)$ is constructible by C4, provided that f_1r is constructible. But f_1r is the composite

$$R(X_1'\ldots A\ldots) \xrightarrow[R(1\ldots h\ldots 1)]{} R(X_1'\ldots Q(B,C)\ldots) \xrightarrow[f_1]{} W_1, \tag{7.21}$$

which is of the form (6.1). The σ of (7.21) is less than (7.20), since $r(Q(W,V_u)) > r(W_1)$. Hence (7.21) is constructible by induction.

This completes the proof.

REFERENCES

[1] G. Gentzen, Untersuchungen über das Logische Schliessen I,II, Math. Z 39(1934-1935), 176-210 and 405-431.

[2] G.M. Kelly, An abstract approach to coherence (in this volume).

[3] G.M. Kelly and S. Mac Lane, Coherence in closed categories, Jour. Pure and Applied Alg. 1(1971), 97-140.

[4] G.M. Kelly and S. Mac Lane, Closed coherence for a natural transformation (in this volume).

[5] J. Lambek, Deductive systems and categories I. Syntactic calculus and residuated categories, Math. Systems theory 2(1968), 287-318.

[6] G. Lewis, Coherence for a closed functor (in this volume).

A NEW RESULT OF COHERENCE FOR DISTRIBUTIVITY

Miguel L. Laplaza

University of Puerto Rico at Mayaguez

Received May 3, 1972

INTRODUCTION

Let $\underline{C}$ be a category with the additional structure given by the following data:

i) Two functors, $\otimes$, $\oplus$: $\underline{C} \times \underline{C} \longrightarrow \underline{C}$.

ii) Two objects, U, N, called the unit and null objects.

iii) For any objects, A, B, C of $\underline{C}$, natural isomorphisms,

$$\alpha_{A,B,C}: A\otimes(B\otimes C) \longrightarrow (A\otimes B)\otimes C, \qquad \gamma_{A,B}: A\otimes B \longrightarrow B\otimes A,$$

$$\alpha'_{A,B,C}: A\oplus(B\oplus C) \longrightarrow (A\oplus B)\oplus C, \qquad \gamma'_{A,B}: A\oplus B \longrightarrow B\oplus A,$$

$$\lambda_A: U\otimes A \longrightarrow A, \qquad \rho_A: A\otimes U \longrightarrow A,$$

$$\lambda'_A: N\oplus A \longrightarrow A, \qquad \rho'_A: A\oplus N \longrightarrow A,$$

$$\lambda^*_A: N\otimes A \longrightarrow N, \qquad \rho^*_A: A\otimes N \longrightarrow N.$$

iv) For any objects, A, B, C of $\underline{C}$, natural monomorphisms,

$$\delta_{A,B,C}: A\otimes(B\oplus C) \longrightarrow (A\otimes B)\oplus(A\otimes C),$$

$$\delta^{\#}_{A,B,C}: (A\oplus B)\otimes C \longrightarrow (A\otimes C)\oplus(B\otimes C).$$

In [5] we have given a coherence theorem for this situation that can be summarized as in [4] or as follows.

Let X be the set $\{x_1, x_2, \cdots, x_p, n, u\}$, $\underline{A}$ the free $\{+, \cdot\}$-algebra over X and $\underline{G}$ the graph consisting of all the following formal symbols for $x, y, z \in \underline{A}$,

$$\alpha_{x,y,z}: x(yz) \longrightarrow (xy)z, \qquad \alpha'_{x,y,z}: x + (y + z) \longrightarrow (x + y) + z,$$

$$\lambda_x: ux \longrightarrow x \quad , \qquad \lambda'_x: n + x \longrightarrow x,$$

$$\rho_x: xu \longrightarrow x \quad , \qquad \rho'_x: x + n \longrightarrow x,$$

$$\gamma_{x,y}: xy \longrightarrow yx \quad , \qquad \gamma'_{x,y}: x + y \longrightarrow y + x,$$

$$\lambda^*_x: nx \longrightarrow n,$$

$$\rho^*_x: xn \longrightarrow n,$$

their formal inverses, indicated by the upper index -1, and,

$$\delta_{x,y,z}: x(y + z) \longrightarrow xy + xz,$$

$$\delta^{\#}_{x,y,z}: (x + y)z \longrightarrow xz + yz,$$

$$1_x: x \longrightarrow x.$$

We construct the free $\{+,\cdot\}$-algebra, $\underline{H}$, over $\underline{G}$ and we take on $\underline{H}$ the unique extension of the graph structure of $\underline{G}$ in which the projections are $\{+,\cdot\}$-morphisms. An element of $\underline{H}$ is an instantiation if, with at most one exception, only elements of $\underline{G}$ of type 1_x are involved in its expression. We denote by $\underline{T}$ the graph consisting of all the instantiations of $\underline{G}$. We defined the paths as the sequences,

$$(*) \qquad y_1 \xrightarrow{\pi_1} y_2 \xrightarrow{\pi_2} \cdots \xrightarrow{\pi_m} y_{m+1} \quad ,$$

where $\pi_i \in \underline{T}$, $i = 1,2,\cdots,m$.

Each map, $f: X \longrightarrow \mathrm{Ob}\ \underline{C}$ such that $f(u) = U$, $f(n) = N$, can be extended in a natural way to a map of the graph $\underline{T}$ onto the graph of the arrows and objects of $\underline{C}$. The value of a path can be defined as the product of the values of the steps and so we have extended the map f to a map of the paths into the arrows of $\underline{C}$ that can be defined as the result of replacing, in each path (*), any occurrence of an element x of X by f(x), every $\cdot$ by $\otimes$ and every + by $\oplus$.

Our paper [5] has studied the conditions on the origin of the path which insure that its value only depends upon the origin and the end of the path. For this we introduced the concept of regularity: let $\underline{A}^*$ be the free $\{+,\cdot\}$-algebra over X with associativity and commutativity for $\cdot$ and +, distributivity of $\cdot$ relatively to +, null element n, identity element u, and the additional condition, $na = an = n$ for $a \in \underline{A}^*$. The identity map of X defines a $\{+,\cdot\}$-morphism, the support, $\mathrm{Supp}:\underline{A} \longrightarrow \underline{A}^*$, and an element x of $\underline{A}$ is regular if Supp(x) can be expressed as a sum of different elements of $\underline{A}^*$ each of which is a product of different elements of X. The coherence result of [5] states that if $P,Q:a \multimap b$ are paths from a to b, $\underline{C}$ is coherent and a is regular, then P and Q have the same value in $\underline{C}$.

Suppose that $\underline{C}$ satisfies the coherence conditions of [5] and that $P,Q:a \multimap b$ are two paths. We want to study the conditions under which P and Q have the same value in $\underline{C}$. We will define for each path a finite sequence of permutations called the distortion and prove that if P and Q have the same distortion, then they have the same value. The method used to define the distortion is the following: we construct a category $\underline{D}$ with a structure of type given by the conditions i) to iv) of this introduction, whose arrows are finite sequences of permutations and define a map, $g:X \longrightarrow \mathrm{Ob}\,\underline{D}$; the value of any path in Arr $\underline{D}$ defined from g is the distortion of the path.

The possibility of a result of the above type was suggested by Saunders Mac Lane in a private communication. The ideas of this paper were inspired largely by the methods used in [1] and [3], although their subsequent development has dimmed that connection. The author is deeply indebted to Professor Mac Lane for his assistance.

§1. Some Preliminary Concepts on Permutations

We are going to explain some concepts that will appear later in the definition of the category of distortions D. We will state also some results intended to ease the proof of the coherence of D: for this we will define a category P, coherent in the sense of [5], such that the coherence theorem of [5] can be used in an effective way to prove some of the required results. These results are contained in the lemmas below; we note that they can be proved in a straightforward (and long) way, taking the definitions independently of the categorical framework. The category P is a full subcategory of Sets with suitable definitions of direct sums and products; it is closely related to the product and permutation categories ("PROPS") of Chapter V of [6]. We will omit all the details in the definitions and proofs that can be completed by a simple routine.

Let P be the full subcategory of Sets whose objects are the sets $\Delta_n = \{1,2,\cdots,n\}$, where n is any natural number. We define two functors + and × by the conditions (for any natural numbers m, m',n,n' and any maps, $f:\Delta_m \longrightarrow \Delta_{m'}$, $g:\Delta_n \longrightarrow \Delta_{n'}$):

i) $\Delta_m + \Delta_n = \Delta_{m+n}$, $\Delta_m \times \Delta_n = \Delta_{mn}$,

ii) $$(f + g)(x) = \begin{cases} f(x), \text{ if } 1 \leq x \leq m, \\ m' + g(x - m), \text{ if } m < x \leq m + n, \end{cases}$$

iii) If $x = i + (j-1)m$, with $1 \leq i \leq m$, $1 \leq j \leq n$, then,
$$(f \times g)(x) = f(i) + [g(j) - 1]m'.$$

It is easy to prove that + and × are associative, with null object $\Delta_0 = \emptyset$ and unit object Δ_1. Moreover × is left distributive relatively to +, and the relation,

$$\Delta_p \times (\Delta_m + \Delta_n) = \Delta_p \times \Delta_m + \Delta_p \times \Delta_n,$$

is natural in Δ_p, Δ_m and Δ_n.

The permutations $t_{p,q}$ and $\tau_{p,q}$ defined by

i) $$t_{p,q}(x) = \begin{cases} q + x, & \text{if } 1 \leq x \leq p, \\ x - p, & \text{if } p < x \leq p + q, \end{cases}$$

ii) if $x = i + (j-1)p$, with $1 \leq i \leq p$, $1 \leq j \leq q$, then,

$$\tau_{p,q}(x) = j + (i-1)q,$$

are natural isomorphisms,

$$t_{p,q} : \Delta_p + \Delta_q \longrightarrow \Delta_q + \Delta_p,$$

$$\tau_{p,q} : \Delta_p \times \Delta_q \longrightarrow \Delta_q \times \Delta_p.$$

We can define now a natural isomorphism,

$$\delta^{\#}_{m,n,p} : (\Delta_m + \Delta_n) \times \Delta_p \longrightarrow (\Delta_m \times \Delta_p) + (\Delta_n \times \Delta_p),$$

by the commutativity of the diagram,

$$\begin{array}{ccc} (\Delta_m + \Delta_n) \times \Delta_p & \xrightarrow{\delta^{\#}_{m,n,p}} & (\Delta_m \times \Delta_p) + (\Delta_n \times \Delta_p) \\ \downarrow \tau_{m+n,p} & & \downarrow \tau_{m,p} + \tau_{n,p} \\ \Delta_p \times (\Delta_m + \Delta_n) & = \!= \!= \!= & (\Delta_p \times \Delta_m) + (\Delta_p \times \Delta_n) \end{array}$$

that is, $\delta^{\#}_{m,n,p} = (\tau_{p,m} + \tau_{p,n})\tau_{m+n,p}$.

If we take for α, α', λ, ρ, λ', ρ', λ^*, ρ^*, δ the identities, and $\gamma_{\Delta_p,\Delta_q} = t_{p,q}$, $\gamma_{\Delta_p,\Delta_q} = \tau_{p,q}$, $\delta^{\#}_{\Delta_p,\Delta_q,\Delta_s} = \delta^{\#}_{p,q,r}$, we have the situation given by the conditions i) to iv) of the Introduction. The proof of the coherence of $\underline{P}$ is easy (see the final remarks of §1 of [5], and observe that the functors + and $\times$ are the direct sum and product).

For all the following lemmas we make the following conventions: P_n is the permutation group of Δ_n, 1_n the identity map of Δ_n, p,q,r and s are any natural numbers, the symbols Δ_p will be represented simply by p when they occur in subindices and the parenthesis and the symbol $\times$ will be omitted when no mistake can arise.

<u>LEMMA 1.1</u>. If $\sigma, \sigma' \in P_p$ and $\tau, \tau' \in P_q$, then,

i) $(\sigma + \tau)(\sigma' + \tau') = \sigma\sigma' + \tau\tau'$,

ii) $(\sigma \times \tau)(\sigma' \times \tau') = \sigma\sigma' \times \tau\tau'$,

iii) $t_{p,q}(\sigma + \tau) = (\tau + \sigma)t_{p,q}$,

iv) $\tau_{p,q}(\sigma \times \tau) = (\tau \times \sigma)\tau_{p,q}$.

The different parts of this lemma are concrete cases of the naturality of $+$, $\times$, $t_{p,q}$ and $\tau_{p,q}$ respectively. The proof is routine.

For the proof of the next lemmas the method is similar; each states the commutativity of a diagram that can be proved by means of the coherence theorem of [5] applied to <u>P</u>. It will be sufficient to show those diagrams in the first three lemmas.

<u>LEMMA 1.2</u>.

i) $t_{p+q,r} = (t_{p,r} + 1_q)(1_p + t_{q,r})$,

ii) $\tau_{pq,r} = (\tau_{p,r} \times 1_q)(1_p \times \tau_{q,r})$.

<u>Proof</u>: The relation i) is the commutativity of the diagram,

$$\begin{array}{ccccc}
\Delta_p + (\Delta_q + \Delta_r) & \xrightarrow{\alpha'_{p,q,r}} & (\Delta_p + \Delta_q) + \Delta_r & \xrightarrow{\gamma'_{p+q,r}} & \Delta_r + (\Delta_p + \Delta_q) \\
\Big\downarrow{\scriptstyle 1_p + \gamma'_{q,r}} & & & & \Big\downarrow{\scriptstyle \alpha'_{r,p,q}} \\
\Delta_p + (\Delta_r + \Delta_q) & \xrightarrow{\alpha'_{p,r,q}} & (\Delta_p + \Delta_r) + \Delta_q & \xrightarrow{\gamma'_{p,r} + 1_q} & (\Delta_r + \Delta_p) + \Delta_q .
\end{array}$$

The diagram for ii) is similar.

<u>LEMMA 1.3</u>.

$$t_{pq,pr} = 1_p \times t_{q,r}.$$

<u>Proof</u>: The lemma states the commutativity of the diagram,

$$\begin{array}{ccc}
\Delta_p \times (\Delta_q + \Delta_r) & \xrightarrow{\delta_{p,q,r}} & \Delta_p \times \Delta_q + \Delta_p \times \Delta_r \\
\Big\downarrow{\scriptstyle 1_p \times \gamma'_{q,r}} & & \Big\downarrow{\scriptstyle \gamma'_{pq,pr}} \\
\Delta_p \times (\Delta_r + \Delta_q) & \xrightarrow{\delta_{p,r,q}} & \Delta_p \times \Delta_r + \Delta_p \times \Delta_q .
\end{array}$$

LEMMA 1.4.

$$(\tau_{r,q} + \tau_{r,p})\tau_{p+q,r}(t_{p,q} \times 1_r) = t_{pr,qr}(\tau_{r,p} + \tau_{r,q})\tau_{p+q,r}.$$

Proof: The lemma states the commutativity of the diagram,

$$\begin{array}{ccccc}
(\Delta_p + \Delta_q)\Delta_r & \xrightarrow[\gamma'_{p,q}\times 1_r]{} & (\Delta_q + \Delta_p)\Delta_r & \xrightarrow[\delta^{\#}_{q,p,r}]{} & \Delta_q\Delta_r + \Delta_p\Delta_r \\
\downarrow \gamma_{p+q,r} & & & & \downarrow \gamma'_{pr,qr} \\
\Delta_r(\Delta_p + \Delta_q) & \xrightarrow{\delta_{r,p,q}} & \Delta_r\Delta_p + \Delta_r\Delta_q & \xrightarrow{\gamma_{r,p}+\gamma_{r,q}} & \Delta_p\Delta_r + \Delta_q\Delta_r
\end{array} .$$

LEMMA 1.5.

$$[1_{ps} + (\tau_{s,q} + \tau_{s,r})\tau_{q+r,s}](\tau_{s,p} + \tau_{s,q+r})$$
$$= [(\tau_{s,p} + \tau_{s,q})\tau_{p+q,s} + 1_{rs}](\tau_{s,p+q} + \tau_{s,r}).$$

LEMMA 1.6.

$$(\tau_{s,pr} + \tau_{s,qr})\tau_{qr+pr,s}[(\tau_{r,p} + \tau_{r,q})\tau_{p+q,r} \times 1_s] = (\tau_{rs,p} + \tau_{rs,q})\tau_{p+q,rs} .$$

LEMMA 1.7.

$$1_p \times [(\tau_{s,q} + \tau_{s,r})\tau_{q+r,s}] = (\tau_{s,pq} + \tau_{s,pr})\tau_{pq+pr,s}.$$

LEMMA 1.8

$$(1_{pr} + t_{pr,qr} + 1_{qs})(\tau_{r+s,p} + \tau_{r+s,q})\tau_{p+q,r+s}$$
$$= (\tau_{r,p} + \tau_{r,q})\tau_{p+q,r} + (\tau_{s,p} + \tau_{s,q})\tau_{p+q,s}.$$

§ 2. The Category of Distortions

Our next aim is to describe the category of distortions, D, which is coherent in the sense of [5]: the distortion of a path will be an arrow of D determined by a method requiring largely the use of the structure of D. We will give the definition of D and its structure but many details are to be omitted: a routine allows one either to check them or reduce the situation to one of the lemmas of §1.

Let $\underline{D}$ be the category whose objects are the finite sequences of natural numbers (the empty sequence ∅ included). We will abbreviate the finite sequence $(a_1, a_2, \cdots, a_r)$ by $(a_\bullet)$ and identify each natural number with the sequence consisting of that number alone. Consequently in the expressions $(a_\bullet)$ and $(b_\bullet)$ it is not supposed that $\bullet$ ranges over the same set of indices. The length of an object of $\underline{D}$ is the usual length of a finite sequence. The set of arrows between sequences of different lengths is empty, and otherwise the definition is given by,

$$\underline{D}[(a_\bullet),(b_\bullet)] = \left\{(\sigma;\ \alpha_1, \cdots, \alpha_r) \mid \sigma \in P_r,\ \alpha_i \in P_{a_i},\ a_i = b_{\sigma(i)}\right\},$$

where r is the common length of $(a_\bullet)$ and $(b_\bullet)$. Hence the existence of an arrow from $(a_\bullet)$ to $(b_\bullet)$ implies that the first sequence can be obtained by a permutation of the second.

The composite of arrows is defined by,

$$\begin{aligned}&(\sigma';\ \alpha'_1, \cdots, \alpha'_r)(\sigma;\ \alpha_1, \cdots, \alpha_r)\\ &= (\sigma'\sigma;\ \alpha'_{\sigma(1)}\alpha_1,\ \alpha'_{\sigma(2)}\alpha_2, \cdots, \alpha'_{\sigma(r)}\alpha_r) .\end{aligned}$$

Thus we have defined a category $\underline{D}$ where the identity of the object $(a_1, \cdots, a_r)$ is $(1_r;\ 1_{a_1}, \cdots, 1_{a_r})$ or simply $(1_r, 1)$ if we make the convention of representing any element of type $(\sigma;\ 1_{n_1}, \cdots, 1_{n_r})$ by $(\sigma;\ 1)$ when no misunderstanding can arise. Note that the identity of the empty sequence is 1_1, the identity map of Δ_1.

The functor $\oplus : \underline{D} \times \underline{D} \longrightarrow \underline{D}$ is defined on the objects by juxtaposition, that is,

$$(a_1, \cdots, a_r) \oplus (b_1, \cdots, b_s) = (a_1, \cdots, a_r, b_1, \cdots, b_s).$$

and on the arrows by using the functor + defined in §1 in the following way:

$$(\sigma;\alpha_1,\cdots,\alpha_r) \oplus (\sigma';\alpha'_1,\cdots,\alpha'_s) = (\sigma + \sigma';\alpha_1,\cdots.\alpha_r,\alpha'_1,\cdots,\alpha'_s).$$

The functor $\oplus$ is associative with null object the empty sequence $\emptyset$. A natural transformation of commutativity,

$$\gamma'_{(a_\bullet),(b_\bullet)} : (a_\bullet) \oplus (b_\bullet) \longrightarrow (b_\bullet) \oplus (a_\bullet) \ ,$$

is defined by, $\gamma'_{(a_\bullet),(b_\bullet)} = (t_{r,s};1)$, where r and s are the lengths of $(a_\bullet)$ and $(b_\bullet)$ respectively.

If we take for α', λ', and ρ' the identities we conclude that $\{\underline{D}, \oplus, \alpha', \gamma', \lambda', \rho', \emptyset\}$ is a coherent situation in the sense of [7] and [2]; that is, $\underline{D}$ is a symmetric monoidal category for $\oplus$ (see [8]).

The functor $\otimes : \underline{D} \times \underline{D} \longrightarrow \underline{D}$ is defined on the objects by,

$$\begin{aligned}(a_1, \cdots, a_r) \otimes (b_1, \cdots, b_s) = (&a_1 + b_1, a_2 + b_1, \cdots, a_r + b_1 ,\\ &a_1 + b_2, a_2 + b_2, \cdots, a_r + b_2 ,\\ &\cdot \quad \cdot \quad \cdot \quad \cdot \quad \cdot \quad \cdot \\ &a_1 + b_s, a_2 + b_s, \cdots, a_r + b_s),\end{aligned}$$

and on the arrows by,

$$\begin{aligned}(\sigma;\ &\alpha_1, \alpha_2, \cdots, \alpha_r) \otimes (\sigma';\ \alpha'_1, \alpha'_2, \cdots, \alpha'_s)\\ &= (\sigma \times \sigma';\ \alpha_1 + \alpha'_1, \alpha_2 + \alpha'_1, \cdots, \alpha_r + \alpha'_1,\\ &\qquad \alpha_1 + \alpha'_2, \alpha_2 + \alpha'_2, \cdots, \alpha_r + \alpha'_2,\\ &\qquad \cdot \quad \cdot \quad \cdot \quad \cdot \quad \cdot \quad \cdot \quad \cdot \\ &\qquad \alpha_1 + \alpha'_s, \alpha_2 + \alpha'_s, \cdots, \alpha_r + \alpha'_s)\end{aligned}$$

with the definition of $\times$ given in §1.

Intuitively the action of $\otimes$ on the objects $(a_\bullet)$ and $(b_\bullet)$ can be described as the result of "reading by columns" the product of the transpose of the matrix $(a_\bullet)$ by $(b_\bullet)$ (using the addition as operation between the elements of the two matrices). A slight modification of this description explains the action of $\otimes$ on a pair of arrows and this proves almost immediately that $\otimes$ is left distributive relatively to $\oplus$.

The functor $\otimes$ is associative with unit object 0 (the sequence with the element 0 only) and both the products $(a_\bullet) \otimes \emptyset$ and $\emptyset \otimes (a_\bullet)$ are the empty sequence $\emptyset$. A natural transformation of commutativity,

$$\gamma_{(a_\bullet),(b_\bullet)} : (a_\bullet) \otimes (b_\bullet) \longrightarrow (b_\bullet) \otimes (a_\bullet) \quad ,$$

is defined by

$$\gamma_{(a_\bullet),(b_\bullet)} = (\tau_{p,q};\ \begin{array}{l} t_{a_1,b_1},\ t_{a_2,b_1},\ \ldots,\ t_{a_p,b_1}\ , \\ t_{a_1,b_2},\ t_{a_2,b_2},\ \ldots,\ t_{a_p,b_2}\ , \\ \cdot \quad\ \ \cdot \quad\ \ \cdot \quad \cdot \quad \cdot \\ t_{a_1,b_s},\ t_{a_2,b_s},\ \ldots,\ t_{a_p,b_s}\,)\ , \end{array}$$

where r and s are the lengths of $(a_\bullet)$ and $(b_\bullet)$ respectively.

If we take for α, λ, ρ the identities we conclude that $\left\{\underline{D},\ \otimes,\ \alpha,\ \gamma,\ \lambda,\ \rho,\ 0\right\}$ is a coherent situation in the sense of [7] and [2]; that is, $\underline{D}$ is a symmetric monoidal category for $\otimes$ (see [8]).

As we have pointed already, $\otimes$ is left distributive relative to $\oplus$ but not right distributive. A natural transformation,

$$\delta^{\#}_{(a_\bullet),(b_\bullet),(c_\bullet)} : [(a_\bullet) \oplus (b_\bullet)] \otimes (c_\bullet) \longrightarrow [(a_\bullet) \otimes (c_\bullet)] \oplus [(b_\bullet) \otimes (c_\bullet)],$$

can be defined by the commutativity of the following diagram, where we have omitted the symbols $\otimes$:

$$\begin{array}{ccc} [(a_\bullet)\oplus(b_\bullet)](c_\bullet) & \xrightarrow[\delta^{\#}_{(a_\bullet),(b_\bullet),(c_\bullet)}]{} & (a_\bullet)(c_\bullet)\oplus(b_\bullet)(c_\bullet) \\ \Big\downarrow{\scriptstyle \gamma_{(a_\bullet)\oplus(b_\bullet),(c_\bullet)}} & & \Big\downarrow{\scriptstyle \gamma_{(a_\bullet),(c_\bullet)}\oplus\gamma_{(b_\bullet),(c_\bullet)}} \\ (c_\bullet)[(a_\bullet)\oplus(b_\bullet)] & = \!= \!= \!= \!= & (c_\bullet)(a_\bullet)\oplus(c_\bullet)(b_\bullet) \end{array}$$

so that we need not check the naturality of the definition. A simple computation proves that if the lengths of $(a_\bullet)$, $(b_\bullet)$ and $(c_\bullet)$ are p, q and r, respectively, then

$$\delta^{\#}_{(a_\bullet),(b_\bullet),(c_\bullet)} = [(\tau_{r,p} + \tau_{r,q})\,\tau_{p+q,r};1] \quad .$$

If we take for δ, λ^* and ρ^* the identities we have the structure given by the conditions i) to iv) of the Introduction and it is easy to prove that $\underline{D}$ is coherent in the sense of [5].

As we have pointed in the Introduction if we fix for each element x_i of $X = \{x_1, x_2, \cdots, x_p, n, u\}$ an object $f(x_i)$ of $\underline{D}$, each path whose steps are instantiations of $\underline{T}$ takes a value that is an arrow of $\underline{D}$. When we take for each $i, f(x_i) = 1$ (the sequence with unique element 1), <u>the value of the path is by definition the distortion of the path.</u> We shall denote by dist(π) the distortion of the path π.

§3. Some Complements on the Construction of the Formal Paths

For a given set, $X = \{x_1, x_2, \cdots, x_p, n, u\}$ we can construct the free $\{+, \cdot\}$- algebra over X, $\underline{A}$, and from it the set of instantiations, $\underline{T}$ (see the Introduction of this paper or §2 of [5]). If we change the set and take $X' = \{x'_1, x'_2, \cdots, x'_{p'}, n', u'\}$ we adopt the convention of denoting the above constructions by $\underline{A}'$ and $\underline{T}'$. Suppose now that $k: X' \longrightarrow X$ is a map such that $k(u') = u$, $k(n') = n$, $\underline{C}$ a category with the structure referred in the Introduction and $f: X \longrightarrow \text{Ob}\,\underline{C}$, $f': X' \longrightarrow \text{Ob}\,\underline{C}$, two maps such that, $f'(u') = f(u) = U$, $f'(n') = f(n) = N$, and that the diagram

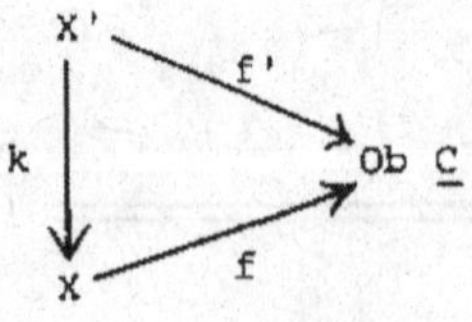

is commutative. The map k can be extended in a natural way to maps from $\underline{A}'$ to $\underline{A}$, from $\underline{T}'$ to $\underline{T}$ and from the paths with steps in $\underline{T}'$ to the paths with steps in $\underline{T}$: the action of these maps can be described intuitively as the result of replacing any occurrence of x_i, u or n by $f(x_i)$, u' or n', respectively. The same symbol k will be used to represent any of these extensions. The maps f and f' define values for the paths with steps in $\underline{T}$ and $\underline{T}'$ respectively and is a simple matter to prove that k preserves the values of the paths; that is, given a path,

$$P: a_1 \xrightarrow{\pi_1} a_2 \xrightarrow{\pi_2} \cdots \xrightarrow{\pi_s} a_{s+1},$$

such that $\pi_i \in \underline{T}$, $1 = 1,2, \cdots, s$, then the value of the path,

$$k(P): k(a_1) \xrightarrow{k(\pi_1)} k(a_2) \xrightarrow{k(\pi_2)} \cdots \xrightarrow{k(\pi_s)} k(a_{s+1}),$$

is the same as the value of P.

If we apply the preceding remarks to the case of the distortion we have proved the following lemma, where X and X' are of the type just described:

LEMMA 3.1. If $k: X' \longrightarrow X$ is a map such that $k(u')=u, k(n')=n$, then for each path P' with steps in T', dist(P') = dist k(P').

LEMMA 3.2. Let X, $\underline{A}$ and $\underline{T}$ be as above and $P, Q: a \xrightarrow{\ominus} b$ two paths whose steps are elements of $\underline{T}$ that are instantiations of α, α', their inverses, γ and γ'. Then it is possible to find a set, $X' = \{x_1', x_2', \cdots, x_p', n', u'\}$, a map $k: X' \longrightarrow X$ and two paths, $P', Q': a' \xrightarrow{\ominus} b'$, with steps in T', such that:

i) $k(u') = u$, $k(n') = n$.

ii) $k(P') = P$, $k(Q') = Q$, $k(a') = a$, $k(b') = b$.

iii) There is no occurrence of the unit element u' in the vertices of the paths P' and Q'.

iv) If in the vertices of P and Q there is no occurrence of n, then in the vertices of P' and Q' there is no occurrence of n'.

<u>Proof</u>: Take $X' = \{x_1, x_2, \cdots, x_p, x_{p-1}, n, u\}$ to be the set obtained adding a new element, x_{p+1}, to X and define $k: X' \longrightarrow X$ taking, $k(x_i) = x_i$ for $i = 1, 2, \cdots, p$, $k(x_{p+1}) = u$, $k(u) = u$ and $k(n) = n$. Then for each element of a $\underline{A}$ there is only one a' such that $k(a') = a$ and no occurrence of the u is in a', and a similar result is true for the elements of $\underline{T}$: given an element, π, of $\underline{T}$ that is an instantiation of α, α', their inverses, γ or γ', there is one and only one element, π', in $\underline{T}'$ such that $k(\pi') = \pi$, with no occurrence of u in the expression of the subscripts of π'. Note that all these elements can be obtained by substituting x_{p+1} for u. The statement of the lemma is now clear.

<u>LEMMA 3.3</u>. Let X and $\underline{A}$ be as above and a an element of $\underline{A}$ with no occurrence of u or n in its expression. Then there exists a set X', a map, $k: X' \longrightarrow X$ and a regular element a' of $\underline{A}'$ such that $k(a') = a$, $k(u') = u$, $k(n') = n$.

<u>Proof</u>: We will construct a' in such a way that the elements of X' occur at most once in the expression of a' and so by Proposition 3 of [5] a' is regular.

The proof of the lemma can be done by induction on $|a|$, norm of a, that is the number of occurrences of elements of X in a (for a formal definition of the norm see §3 of [5]). If $a = a_1 + a_2$, construct by the induction hypothesis the sets X_1' and X_2', the maps $k_1: X_1' \longrightarrow X$, $k_2: X_2' \longrightarrow X$ such that, $k_1(a_1') = a_1$, $k(a_2') = a_2$, where a_1' and a_2' are regular elements; we can suppose that $X_1' \cap X_2' = \{u, n\}$, unit and null elements of X, and we take $X' = X_1' \cup X_2'$ and for k the extension of k_1 and k_2 to X'. If $a = a_1 a_2$, a similar construction can be used.

<u>LEMMA 3.4</u>. Let $X, X', k: X' \longrightarrow X$, $\underline{A}$, $\underline{A}'$, $\underline{T}$ and $\underline{T}'$ be as in the beginning of §3, $\pi: a \longrightarrow b$ an element of $\underline{T}$ that is an instantiation of α, α', their inverses, γ or γ', and a' an element of $\underline{A}'$ such that $k(a') = a$. Then there exists an element π' of $\underline{T}'$ such that $k(\pi') = \pi$.

<u>Proof</u>: The proof can be done by induction on $|a|$. If π is an element of type α, α', their inverses, γ or γ', then it is easy to do; we will give details only in the case $\pi = \alpha_{x,y,z}$: then $a = x(yz)$, and this implies that, $a' = x'(y'z')$, $k(x') = x$, $k(y') = y$, $k(z') = z$, and it is sufficient to take $\pi' = \alpha_{x',y',z'}$.

Suppose now that $\pi = \pi_1 + \pi_2$, $\pi_1 : a_1 \longrightarrow b_1$, $\pi_2 : a_2 \longrightarrow b_2$. Then $a = a_1 + a_2$, and this implies that, $a' = a_1' + a_2'$, $k(a_1') = a_1$, $k(a_2') = a_2$; using the induction hypothesis we can determine π_1' and π_2' such that, $k(\pi_1') = \pi_1$, $k(\pi_2') = \pi_2$, and we can take $\pi' = \pi_1' + \pi'_2$. The case $\pi = \pi_1\pi_2$ is similar.

<u>LEMMA 3.5</u>. Let X, <u>A</u> and <u>T</u> be as before and P a path with steps in <u>T</u>,

$$P : a \xrightarrow{\pi_1} a_1 \xrightarrow{\pi_2} \cdots \xrightarrow{\pi_n} a_n ,$$

that are instantiations of α, α', their inverses, γ, and γ'. If no occurrence of u and n is in the expression of an element a of <u>A</u>, then there exists a set X', a regular element a' of X', a map $k : X' \longrightarrow X$ and a path,

$$P' : a' \xrightarrow{\pi_1} a'_1 \xrightarrow{\pi_2} \cdots \xrightarrow{\pi_n} a_n' ,$$

such that, $k(u') = u$, $k(n') = n$, $k(a') = a$, $k(a_i') = a_i$ and $k(\pi_i') = \pi_i$ for $i = 1, 2, \cdots, n$. Moreover the choice of X', k and a' only depends upon a, independently of the other components of the path P.

<u>Proof</u>: This is an immediate consequence of Lemmas 3.3 and 3.4.

Note that if $P, Q : a \dashrightarrow b$ are two paths allowing us to apply Lemma 3.5 we will find two paths, $P' : a' \dashrightarrow b'$, $Q' : a' \dashrightarrow b''$, where, in general, $b' \neq b''$. Our next aim is to find conditions under which we can deduce that $b' = b''$; as we will see, this is the case when P and Q have the same distortion.

We now need an auxiliary concept. Let <u>S</u> be the free $\{+, \cdot\}$-algebra over $\{*\}$, the set with only one element. <u>The shape</u> is the map, $sh : \underline{A} \longrightarrow S$ defined by the conditions:

i) For $x \in X$, $sh(x) = *$.

ii) $x = a + b \Longrightarrow sh(x) = xh(a) + sh(b)$.

iii) $x = ab \Longrightarrow sh(x) = sh(a)sh(b)$.

It is immediate that, taking X, X' and $k:X' \longrightarrow X$ as before, we have for every a' of $\underline{A}'$, $sh\, k(a') = sh(a')$.

<u>LEMMA 3.6</u>. Let X, $\underline{A}$ and $\underline{T}$ be as before and suppose that $P:a \xrightarrow{\Theta} b$, $P':a \xrightarrow{\Theta} b'$ are paths with steps in $\underline{T}$ such that

i) a is a sum of elements that are products of elements of X,

ii) $sh(b) = sh(b')$,

iii) $dist(P) = dist(P')$,

iv) P and P' are sequences of identities and instantiations of α, α', their inverses, γ and γ',

then, $b = b'$.

<u>Proof</u>: Note that using the definitions of §3 of [5], the relation $sh(b) = sh(b')$ implies that b and b' have the same number of additive or multiplicative components and if, $Adec\, b = (b_1, b_2, \cdots, b_r)$, $Adec\, b' = (b'_1, b'_2, \cdots, b'_r)$, then b_i and b'_i have the same number of multiplicative components. Moreover, $Apt\, b = Apt\, b'$ and $Mpt\, b = Mpt\, b'$. All this can be proved immediately by induction on $|b|$.

Suppose that for each i, $Mdec\, b_i = (b_{i1}, b_{i2}, \cdots, b_{in_i})$, $Mdec\, b'_i = (b'_{i1}, b'_{i2}, \cdots, b'_{in_i})$. If we prove that for every pair $<i,j>$, $b_{ij} = b'_{ij}$, using Proposition 1 of [5] we have that $b_i = b'_i$ for $i = 1, 2, \cdots, r$, and applying again that proposition we have that $b = b'$.

To prove that $b_{ij} = b'_{ij}$ it will be sufficient to show that b_{ij} can be computed by means of a and dist(P), and in fact we are going to prove that if $Adec\, a = (a_1, a_2, \cdots, a_r)$, $Mdec\, a_i = (a_{i1}, a_{i2}, \cdots, a_{im_i})$ and $dist(P) = (\sigma; \alpha_1, \alpha_2, \cdots, \alpha_r)$, then $m_j = n_{\sigma(j)}$ and $a_{i,j} = b_{\sigma(i), \alpha_i(j)}$. Note that a and b have the same number of additive components as a consequence of condition iv). So we are reduced to proving the above statement and we are going to do this by induction on the number of steps in the path P. Suppose that,

$$P = a \xrightarrow{\;Q\;} c \xrightarrow{\;R\;} b,$$

where, Adec $c = (c_1, c_2, \cdots, c_r)$, Mdec $c_i = (c_{i1}, c_{i2}, \cdots, c_{in_i})$,

$$\text{dist}(Q) = (\tau; \beta_1, \cdots, \beta_r),\ \text{dist}(R) = (\omega; \gamma_1, \cdots, \gamma_r),$$

then,

$$(\sigma;\ \alpha_i) = \text{dist}(P) = \text{dist}(R)\text{dist}(Q) = (\omega;\ \gamma_i)(\tau;\beta_i)$$
$$= (\omega\tau;\ \gamma_{\tau(i)}\beta_i),$$

that is, $\sigma = \omega\tau$, $\alpha_i = \gamma_{\tau(i)}\beta_i$ for $i = 1,2,\cdots,r$, and by the induction hypothesis,

$$a_{ij} = c_{\tau(i),\beta_i(j)} = b_{\omega[\tau(i)],\ \gamma_{\tau(i)}[\beta_i(j)]} = b_{\sigma(i),\ \alpha_i(j)}.$$

Hence we are reduced to the case when P has only one step, that is, when $P = \pi : a \longrightarrow b$ is an element of $\underline{T}$ and this will be done by induction on $|a|$. If $\pi = \pi' + \pi''$, then we can suppose, $a = a' + a''$, $b = b' + b''$, $\pi' : a' \longrightarrow b'$, $\pi'' : a'' \longrightarrow b''$,

$$\text{dist}(\pi') = (\sigma';\ \alpha'_1, \cdots, \alpha'_p),\ \text{dist}(\pi'') = (\sigma''; \alpha''_1, \cdots, \alpha''_q)\ ,$$
$$\text{Adec } a' = (a_1, a_2, \cdots, a_p),\ \text{Adec } a'' = (a_{p+1}, a_{p+2}, \cdots, a_{p+q})\ ,$$
$$\text{Adec } b' = (b_1, b_2, \cdots, b_p),\ \text{Adec } b'' = (b_{p+1}, b_{p+2}, \cdots, b_{p+q})\ ,$$

and by the induction hypothesis, if $i \leq p$,

$$a_{ij} = a'_{ij} = b'_{\sigma'(i),\alpha'_i(j)} = b_{\sigma'(i),\alpha'_i(j)}\ ,$$

and for $i > p$,

$$a_{ij} = a''_{i-p,j} = b''_{\sigma''(i-p),\alpha''_{i-p}(j)} = b_{p+\sigma''(i-p),\alpha''_{i-p}(j)}\ .$$

Hence, $\text{dist}(\pi) = \text{dist}(\pi') \oplus \text{dist}\ (\pi'') = (\sigma' + \sigma''; \alpha'_1, \alpha'_2, \cdots$
$\cdots, \alpha'_p, \alpha''_1, \alpha''_2, \cdots, \alpha''_q) = (\sigma;\ \alpha_1, \alpha_2, \cdots, \alpha_{p+q})$,
and for $i \leq p$,

$$a_{ij} = b_{\sigma'(i),\alpha_i'(j)} = b_{(\sigma'+\sigma'')(i),\alpha_i'(j)} = b_{\sigma(i),\alpha_i(j)} \; ,$$

and for $i > p$,

$$a_{ij} = b_{p+\sigma''(i-p),\alpha''_{i-p}(j)} = b_{(\sigma'+\sigma'')(i),\alpha''_{i-p}(j)} = b_{\sigma(i),\alpha_i(j)} \; .$$

If $\pi = \pi' \pi''$, $a = a'a''$, $b = b'b''$, $\pi':a' \longrightarrow b'$, $\pi'':a'' \longrightarrow b''$, then, taking into account condition iv),

$$\mathrm{dist}(\pi') = (1;\alpha'), \; \mathrm{dist}(\pi'') = (1;\alpha'') \; ,$$

$$(1;\alpha) = \mathrm{dist}(\pi) = (1;\alpha' + \alpha'') \; ,$$

$$\mathrm{Adec}\; a = (a_1) = (a), \; \mathrm{Adec}\; a' = (a_1') = (a') \; ,$$

$$\mathrm{Adec}\; a'' = (a_1'') = (a''), \; \mathrm{Mdec}\; a_1 = (a_{11}, a_{12}, \cdots a_{1n}) \; ,$$

$$\mathrm{Mdec}\; a' = (a_{11}, \cdots, a_{1p}), \; \mathrm{Mdec}\; a'' = (a_{1,p+1}, \cdots, a_{1n}) \; ,$$

$$\mathrm{Adec}\; b = (b_1) = (b), \; \mathrm{Mdec}\; b_1 = (b_{11}, \cdots, b_{1n}) \; ,$$

and by the induction hypothesis, if $i \le p$, $a_{1i} = b_{1,\alpha'(i)}$, and for $i > p$, $a_{1i} = b_{1,p+\alpha''(1-p)}$. From the above remarks it follows that, $\alpha = \alpha' + \alpha''$ and for $i \le p$,

$$a_{1i} = b_{1,\alpha'(i)} = b_{1,\alpha(i)} \; ,$$

and for $i > p$,

$$a_{1i} = b_{1,p+\alpha''(i-p)} = b_{1,\alpha(i)} \; .$$

So we are reduced to the case where π is an element of $\underline{G}$, that is an identity or an element of type α, α', their inverses, γ or γ'. If π is of type α, α' or their inverses, $\mathrm{dist}(\pi)$ is an identity and the result is immediate. Suppose now that $\pi = \gamma_{a',a''}:a'a'' \longrightarrow a''a'$; then by condition iv),

$$\text{Adec } a = (a) = (a_1),\ \text{Mdec } a_1 = (a_{11}, a_{12}, \cdots, a_{1n}) \quad ,$$

$$\text{Adec } a' = (a'),\ \text{Mdec } a' = (a_{11}, a_{12}, \cdots, a_{1p}),$$

$$\text{Adec } a'' = (a''),\ \text{Mdec } a'' = (a_{1,p+1}, \cdots a_{1n}) \quad .$$

Then, $\text{dist}(\pi) = (\tau_{1,1}; t_{p,n-p}) = (1_1; t_{p,n-p})$,

$$\text{Adec } b = (b),\ \text{Mdec } b = (a_{1,p+1}, \cdots, a_r, a_1, \cdots, a_p)$$

and the relation can be checked immediately.

The only remaining case is when $\pi = \gamma'_{a',a''}: a' + a'' \longrightarrow a'' + a'$. If,

$$\text{Adec } a = (a_1, a_2, \cdots, a_r),\ \text{Mdec } a_i = (a_{i1}, a_{i2}, \cdots, a_{in_i}) \quad ,$$

$$\text{Adec } a' = (a_1, a_2, \cdots, a_p),\ \text{Adec } a'' = (a_{p+1}, a_{p+2}, \cdots, a_r) \ ,$$

we have,

$$\text{dist}(\pi) = (t_{p,r-p}; 1),\ \text{Adec } b = (a_{p+1}, \cdots, a_r, a_1, a_2, \cdots, a_p) \ ,$$

and the relation can be checked immediately.

§4. The Result of Coherence

We are now going to prove our coherence result and for this we will suppose that $\underline{C}$ is the category referred to in the Introduction, where the concepts represented by X, $\underline{A}$ and $\underline{T}$ are also defined. We will suppose that X and $f: X \longrightarrow \text{Ob } \underline{C}$ are fixed, so that any path whose construction is not detailed is defined over $\underline{T}$ and takes its value on Ob $\underline{C}$ by means of f.

COHERENCE THEOREM

If $\underline{C}$ satisfies the coherence conditions of [5] and $P, Q: a \dashrightarrow b$ are paths with the same distortion, then P and Q take the same value in the category $\underline{C}$.

Proof: The proof will consist of different parts that reduce the general situation of the paths P and Q to others easier to handle.

Part I: We reduce the theorem to the case where no occurrence of n is in the vertices of P and Q.

Take the definition of reduction of §4 of [5]: a reduction of an element d of $\underline{A}$ is a path, $d \xrightarrow{\ominus} d'$, obtained applying as many times as possible instantiations of $\lambda', \rho', \lambda^*$ and ρ^*; roughly speaking a reduction is a path obtained by successive elimination of the occurrences of n. The value and the end of a reduction, $d \xrightarrow{\ominus} d'$, are determined by d (Proposition 4 of [5])and according to Proposition 5 of [5] we can construct two commutative diagrams,

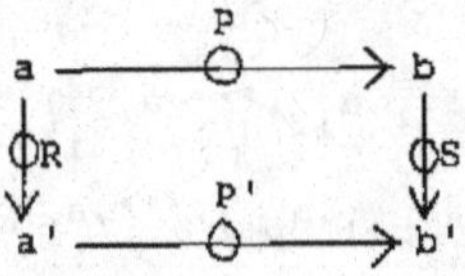

,
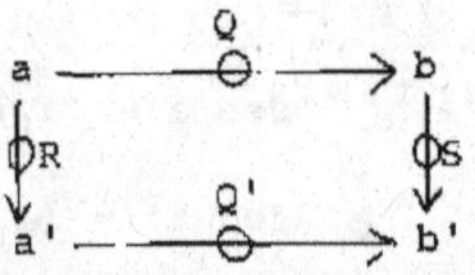

,

where R and S are reductions of a and b, respectively, and no instantiation of λ', ρ', λ^*, ρ^* or their inverses is in P' or Q'. If some occurrence of n is in a', then a' = n, a' is regular and the theorem is consequence of Proposition 10 of [5]: hence we can suppose that no occurrence of n is in any of the vertices of P' and Q'. The commutativity of the above diagrams imply the commutativity of the diagrams obtained by replacing the paths by their values or by their distortions, and as the value of any path is a monic of $\underline{C}$ and the distortion is an isomorphism of $\underline{D}$ (the category of distortions), we can conclude that, dist(P') = dist(Q') and that we are reduced to proving that P' and Q' take the same value in $\underline{C}$.

The same type of argument will be used in the next two Parts and some details can (and will) be omitted.

Part II: We reduce the theorem to the case with the additional conditions that all the vertices of P and Q are sums of products of elements of X and that no instantiation of δ or $\delta^{\#}$ is in P or Q.

A rappel of an element d of $\underline{A}$ is a path $d \xrightarrow{\ominus} d'$ obtained by applying as many times as possible instantiations of δ and $\delta^{\#}$ (see

§ 5 of [5]): the end of any rappel has to be the sum of products of elements of X. Proposition 7 of [5] proves the existence of two commutative diagrams of type,

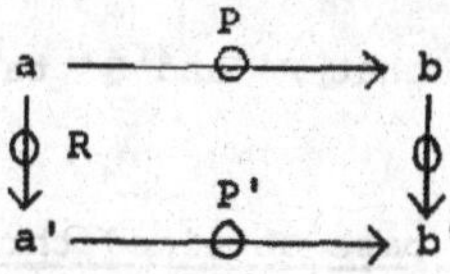

,
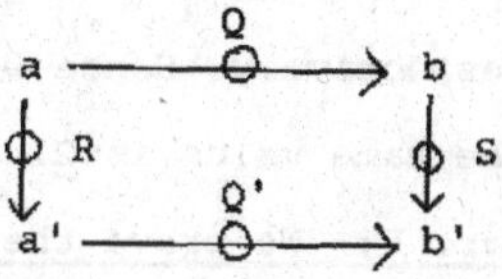

,

where R and S are rappels of a and b, respectively, and no instantiation of λ', ρ', λ^*, ρ^*, their inverses, δ or $\delta^{\#}$ is in P' or Q'. As in Part I, dist(P')= dist(Q') and we are reduced to proving that P' and Q' take the same value in $\underline{C}$.

<u>Part III: We reduce the theorem to the case with the additional hypothesis that no instantiation of λ or ρ is in P or Q.</u>

A normalization of an element of $\underline{A}$ is a path $d \overset{\ominus}{\longrightarrow} d'$ obtained by applying as many times as possible instantiations of λ and ρ (see § 6 of [5]). The value and the end of a normalization, $d \overset{\ominus}{\longrightarrow} d'$, are determined by d(Proposition 8 of [5]). Proposition 9 of [5] proves the existence of two commutative diagrams of type,

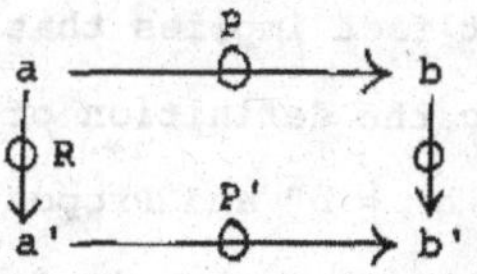

,
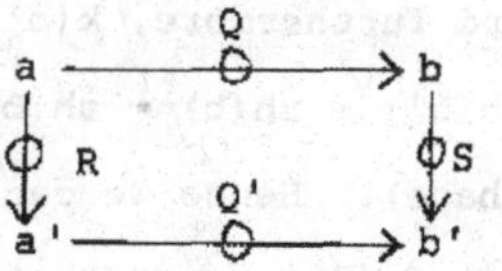

,

where R and S are normalizations of a and b, respectively, and P' and Q' are instantiations of α, α', their inverses, γ and γ'. A consequence of the above facts is that dist(P') = dist(Q') and that we are reduced to proving that P' and Q; take the same value in $\underline{C}$.

<u>Part IV: We reduce the theorem to the case with the additional condition that no occurrence of u is in the vertices of P and $\underline{Q}$.</u>

We can use Lemma 3.2 to construct a set X', a map $k:X' \longrightarrow X$ and two paths, $P', Q':a' \overset{\ominus}{\longrightarrow} b'$ such that k(P') = P, k(Q') = Q, with no occurrence of u' in the vertices of P' and Q'. By Lemma 3.1,

$$\mathrm{dist}(P') = \mathrm{dist}(P) = \mathrm{dist}(Q) = \mathrm{dist}(Q').$$

Define now $f':X' \longrightarrow \mathrm{Ob}\ \underline{C}$ by the composition, $X' \xrightarrow{k} X \xrightarrow{f} \mathrm{Ob}\ \underline{C}$; then the values of P and Q are the same as the values of P' and Q', respectively. Hence we are reduced to proving that P' and Q' take the same value in $\underline{C}$.

Part V: We prove the theorem for the reduced case obtained through Parts I to IV.

We are reduced to proving the theorem for two paths P,Q that are sequences of identities and instantiations of α, α', their inverses, γ and γ', and whose vertices are sums of products of elements of X with no occurrence of u and n. By Lemma 3.5 we can construct a set X', a map $k:X' \longrightarrow X$, two paths, $P':a' \xrightarrow{\Theta} b'$, $Q':a' \xrightarrow{\Theta} b''$ such that, $k(P') = P$, $k(Q') = Q$, where a' is a regular element of $\underline{A}'$. Define f' as in Part IV and we are reduced to proving that P' and Q' take the same value in $\underline{C}$; $k(P') = P$ and $k(Q') = Q$, imply by Lemma 3.1 that,

$$\mathrm{dist}(P') = \mathrm{dist}(P) = \mathrm{dist}(Q) = \mathrm{dist}(Q'),$$

and furthermore, $k(b') = k(b'') = b$. This last fact implies that $\mathrm{sh}(b') = \mathrm{sh}(b) = \mathrm{sh}(b'')$ (see remarks following the definition of shape). Hence we can use Lemma 3.6 to obtain $b' = b''$ and Proposition 10 of [5] proves that $P', Q':a' \xrightarrow{\Theta} b'$ take the same value in $\underline{C}$ and the theorem is proved.

REFERENCES

[1] S. Eilenberg and G.M. Kelly: "A generalization of the functorial calculus", J. Algebra, 3(1966), 366-375.

[2] G.M. Kelly: "On Mac Lane's conditions for coherence of natural associativities, commutativities, etc.", J.Algebra, 4(1964), 397-402.

[3] G.M. Kelly and S. Mac Lane: "Coherence in closed categories", J. Pure Appl. Algebra, 1(1971), 97-140.

[4] M.L. Laplaza: "Coherence for categories with associativity, commutativity and distributivity", Bull. Amer. Math. Soc., 78 (1972), 220-222.

[5] M.L. Laplaza: "Coherence for distributivity", (this volume).

[6] S. Mac Lane: "Categorical algebra", Bull. Amer. Math. Soc., 71 (1965), 40-106.

[7] S. Mac Lane: "Natural associativity and commutativity", Rice Univ. Studies, 49(1963), No. 4, 28-46.

[8] S. Mac Lane: "Categories for the working mathematician", New York-Heidelberg-Berlin, Springer, 1971.

Lecture Notes in Mathematics

Comprehensive leaflet on request

Vol. 74: A. Fröhlich, Formal Groups. IV, 140 pages. 1968. DM 16,-

Vol. 75: G. Lumer, Algèbres de fonctions et espaces de Hardy. VI, 80 pages. 1968. DM 16,-

Vol. 76: R. G. Swan, Algebraic K-Theory. IV, 262 pages. 1968. DM 18,-

Vol. 77: P.-A. Meyer, Processus de Markov: la frontière de Martin. IV, 123 pages. 1968. DM 16,-

Vol. 78: H. Herrlich, Topologische Reflexionen und Coreflexionen. XVI, 166 Seiten. 1968. DM 16,-

Vol. 79: A. Grothendieck, Catégories Cofibrées Additives et Complexe Cotangent Relatif. IV, 167 pages. 1968. DM 16,-

Vol. 80: Seminar on Triples and Categorical Homology Theory. Edited by B. Eckmann. IV, 398 pages. 1969. DM 20,-

Vol. 81: J.-P. Eckmann et M. Guenin, Méthodes Algébriques en Mécanique Statistique. VI, 131 pages. 1969. DM 16,-

Vol. 82: J. Wloka, Grundräume und verallgemeinerte Funktionen. VIII, 131 Seiten. 1969. DM 16,-

Vol. 83: O. Zariski, An Introduction to the Theory of Algebraic Surfaces. Second Printing. IV, 100 pages. 1972. DM 16,-

Vol. 84: H. Lüneburg, Transitive Erweiterungen endlicher Permutationsgruppen. IV, 119 Seiten. 1969. DM 16,-

Vol. 85: P. Cartier et D. Foata, Problèmes combinatoires de commutation et réarrangements. IV, 88 pages. 1969. DM 16,-

Vol. 86: Category Theory, Homology Theory and their Applications I. Edited by P. Hilton. VI, 216 pages. 1969. DM 16,-

Vol. 87: M. Tierney, Categorical Constructions in Stable Homotopy Theory. IV, 65 pages. 1969. DM 16,-

Vol. 88: Séminaire de Probabilités III. IV, 229 pages. 1969. DM 18,-

Vol. 89: Probability and Information Theory. Edited by M. Behara, K. Krickeberg and J. Wolfowitz. IV, 256 pages. 1969. DM 18,-

Vol. 90: N. P. Bhatia and O. Hajek, Local Semi-Dynamical Systems. II, 157 pages. 1969. DM 16,-

Vol. 91: N. N. Janenko, Die Zwischenschrittmethode zur Lösung mehrdimensionaler Probleme der mathematischen Physik. VIII, 194 Seiten. 1969. DM 16,80

Vol. 92: Category Theory, Homology Theory and their Applications II. Edited by P. Hilton. V, 308 pages. 1969. DM 20,-

Vol. 93: K. R. Parthasarathy, Multipliers on Locally Compact Groups. III, 54 pages. 1969. DM 16,-

Vol. 94: M. Machover and J. Hirschfeld, Lectures on Non-Standard Analysis. VI, 79 pages. 1969. DM 16,-

Vol. 95: A. S. Troelstra, Principles of Intuitionism. II, 111 pages. 1969. DM 16,-

Vol. 96: H.-B. Brinkmann und D. Puppe, Abelsche und exakte Kategorien, Korrespondenzen. V, 141 Seiten. 1969. DM 16,-

Vol. 97: S. O. Chase and M. E. Sweedler, Hopf Algebras and Galois theory. II, 133 pages. 1969. DM 16,-

Vol. 98: M. Heins, Hardy Classes on Riemann Surfaces. III, 106 pages. 1969. DM 16,-

Vol. 99: Category Theory, Homology Theory and their Applications III. Edited by P. Hilton. IV, 489 pages. 1969. DM 24,-

Vol. 100: M. Artin and B. Mazur, Etale Homotopy. II, 196 Seiten. 1969. DM 16,-

Vol. 101: G. P. Szegö et G. Treccani, Semigruppi di Trasformazioni Multivoche. VI, 177 pages. 1969. DM 16,-

Vol. 102: F. Stummel, Rand- und Eigenwertaufgaben in Sobolewschen Räumen. VIII, 386 Seiten. 1969. DM 20,-

Vol. 103: Lectures in Modern Analysis and Applications I. Edited by C. T. Taam. VII, 162 pages. 1969. DM 16,-

Vol. 104: G. H. Pimbley, Jr., Eigenfunction Branches of Nonlinear Operators and their Bifurcations. II, 128 pages. 1969. DM 16,-

Vol. 105: R. Larsen, The Multiplier Problem. VII, 284 pages. 1969. DM 18,-

Vol. 106: Reports of the Midwest Category Seminar III. Edited by S. Mac Lane. III, 247 pages. 1969. DM 16,-

Vol. 107: A. Peyerimhoff, Lectures on Summability. III, 111 pages. 1969. DM 16,-

Vol. 108: Algebraic K-Theory and its Geometric Applications. Edited by R. M. F. Moss and C. B. Thomas. IV, 86 pages. 1969. DM 16,-

Vol. 109: Conference on the Numerical Solution of Differential Equations. Edited by J. Ll. Morris. VI, 275 pages. 1969. DM 18,-

Vol. 110: The Many Facets of Graph Theory. Edited by G. Chartrand and S. F. Kapoor. VIII, 290 pages. 1969. DM 18,-

Vol. 111: K. H. Mayer, Relationen zwischen charakteristischen Zahlen. III, 99 Seiten. 1969. DM 16,-

Vol. 112: Colloquium on Methods of Optimization. Edited by N. N. Moiseev. IV, 293 pages. 1970. DM 18,-

Vol. 113: R. Wille, Kongruenzklassengeometrien. III, 99 Seiten. 1970. DM 16,-

Vol. 114: H. Jacquet and R. P. Langlands, Automorphic Forms on GL (2). VII, 548 pages. 1970. DM 24,-

Vol. 115: K. H. Roggenkamp and V. Huber-Dyson, Lattices over Orders I. XIX, 290 pages. 1970. DM 18,-

Vol. 116: Séminaire Pierre Lelong (Analyse) Année 1969. IV, 195 pages. 1970. DM 16,-

Vol. 117: Y. Meyer, Nombres de Pisot, Nombres de Salem et Analyse Harmonique. 63 pages. 1970. DM 16,-

Vol. 118: Proceedings of the 15th Scandinavian Congress, Oslo 1968. Edited by K. E. Aubert and W. Ljunggren. IV, 162 pages. 1970. DM 16,-

Vol. 119: M. Raynaud, Faisceaux amples sur les schémas en groupes et les espaces homogènes. III, 219 pages. 1970. DM 16,-

Vol. 120: D. Siefkes, Büchi's Monadic Second Order Successor Arithmetic. XII, 130 Seiten. 1970. DM 16,-

Vol. 121: H. S. Bear, Lectures on Gleason Parts. III, 47 pages. 1970. DM 16,-

Vol. 122: H. Zieschang, E. Vogt und H.-D. Coldewey, Flächen und ebene diskontinuierliche Gruppen. VIII, 203 Seiten. 1970. DM 16,-

Vol. 123: A. V. Jategaonkar, Left Principal Ideal Rings. VI, 145 pages. 1970. DM 16,-

Vol. 124: Séminare de Probabilités IV. Edited by P. A. Meyer. IV, 282 pages. 1970. DM 20,-

Vol. 125: Symposium on Automatic Demonstration. V, 310 pages. 1970. DM 20,-

Vol. 126: P. Schapira, Théorie des Hyperfonctions. XI, 157 pages. 1970. DM 16,-

Vol. 127: I. Stewart, Lie Algebras. IV, 97 pages. 1970. DM 16,-

Vol. 128: M. Takesaki, Tomita's Theory of Modular Hilbert Algebras and its Applications. II, 123 pages. 1970. DM 16,-

Vol. 129: K. H. Hofmann, The Duality of Compact Semigroups and C*-Bigebras. XII, 142 pages. 1970. DM 16,-

Vol. 130: F. Lorenz, Quadratische Formen über Körpern. II, 77 Seiten. 1970. DM 16,-

Vol. 131: A. Borel et al., Seminar on Algebraic Groups and Related Finite Groups. VII, 321 pages. 1970. DM 22,-

Vol. 132: Symposium on Optimization. III, 348 pages. 1970. DM 22,-

Vol. 133: F. Topsøe, Topology and Measure. XIV, 79 pages. 1970. DM 16,-

Vol. 134: L. Smith, Lectures on the Eilenberg-Moore Spectral Sequence. VII, 142 pages. 1970. DM 16,-

Vol. 135: W. Stoll, Value Distribution of Holomorphic Maps into Compact Complex Manifolds. II, 267 pages. 1970. DM 18,-

Vol. 136: M. Karoubi et al., Séminaire Heidelberg-Saarbrücken-Strasbourg sur la K-Théorie. IV, 264 pages. 1970. DM 18,-

Vol. 137: Reports of the Midwest Category Seminar IV. Edited by S. MacLane. III, 139 pages. 1970. DM 16,-

Vol. 138: D. Foata et M. Schützenberger, Théorie Géométrique des Polynômes Eulériens. V, 94 pages. 1970. DM 16,-

Vol. 139: A. Badrikian, Séminaire sur les Fonctions Aléatoires Linéaires et les Mesures Cylindriques. VII, 221 pages. 1970. DM 18,-

Vol. 140: Lectures in Modern Analysis and Applications II. Edited by C. T. Taam. VI, 119 pages. 1970. DM 16,-

Vol. 141: G. Jameson, Ordered Linear Spaces. XV, 194 pages. 1970. DM 16,-

Vol. 142: K. W. Roggenkamp, Lattices over Orders II. V, 388 pages. 1970. DM 22,-

Vol. 143: K. W. Gruenberg, Cohomological Topics in Group Theory. XIV, 275 pages. 1970. DM 20,-

Vol. 144: Seminar on Differential Equations and Dynamical Systems, II. Edited by J. A. Yorke. VIII, 268 pages. 1970. DM 20,-

Vol. 145: E. J. Dubuc, Kan Extensions in Enriched Category Theory. XVI, 173 pages. 1970. DM 16,-

Please turn over

Vol. 146: A. B. Altman and S. Kleiman, Introduction to Grothendieck Duality Theory. II, 192 pages. 1970. DM 18,–

Vol. 147: D. E. Dobbs, Cech Cohomological Dimensions for Commutative Rings. VI, 176 pages. 1970. DM 16,–

Vol. 148: R. Azencott, Espaces de Poisson des Groupes Localement Compacts. IX, 141 pages. 1970. DM 16,–

Vol. 149: R. G. Swan and E. G. Evans, K-Theory of Finite Groups and Orders. IV, 237 pages. 1970. DM 20,–

Vol. 150: Heyer, Dualität lokalkompakter Gruppen. XIII, 372 Seiten. 1970. DM 20,–

Vol. 151: M. Demazure et A. Grothendieck, Schémas en Groupes I. (SGA 3). XV, 562 pages. 1970. DM 24,–

Vol. 152: M. Demazure et A. Grothendieck, Schémas en Groupes II. (SGA 3). IX, 654 pages. 1970. DM 24,–

Vol. 153: M. Demazure et A Grothendieck, Schémas en Groupes III. (SGA 3). VIII, 529 pages 1970. DM 24,–

Vol. 154: A. Lascoux et M. Berger, Variétés Kähleriennes Compactes. VII, 83 pages. 1970. DM 16,–

Vol. 155: Several Complex Variables I, Maryland 1970. Edited by J. Horváth. IV, 214 pages. 1970. DM 18,–

Vol. 156: R. Hartshorne, Ample Subvarieties of Algebraic Varieties. XIV, 256 pages. 1970. DM 20,–

Vol. 157: T. tom Dieck, K. H. Kamps und D. Puppe, Homotopietheorie. VI, 265 Seiten. 1970. DM 20,–

Vol. 158: T. G. Ostrom, Finite Translation Planes. IV. 112 pages. 1970. DM 16,–

Vol. 159: R. Ansorge und R. Hass, Konvergenz von Differenzenverfahren für lineare und nichtlineare Anfangswertaufgaben. VIII, 145 Seiten. 1970 DM 16,–

Vol. 160: L. Sucheston, Constributions to Ergodic Theory and Probability. VII, 277 pages. 1970. DM 20,–

Vol. 161: J Stasheff, H-Spaces from a Homotopy Point of View. VI, 95 pages. 1970. DM 16,–

Vol. 162: Harish-Chandra and van Dijk, Harmonic Analysis on Reductive p-adic Groups. IV, 125 pages. 1970. DM 16,–

Vol. 163: P. Deligne, Equations Différentielles à Points Singuliers Reguliers. III, 133 pages. 1970. DM 16,–

Vol. 164: J. P. Ferrier, Seminaire sur les Algebres Complètes. II, 69 pages. 1970. DM 16,–

Vol. 165: J. M. Cohen, Stable Homotopy V, 194 pages 1970 DM 16,–

Vol. 166: A. J. Silberger, PGL_2 over the p-adics: its Representations, Spherical Functions, and Fourier Analysis. VII, 202 pages. 1970. DM 18,–

Vol. 167: Lavrentiev, Romanov and Vasiliev, Multidimensional Inverse Problems for Differential Equations. V, 59 pages. 1970 DM 16,–

Vol. 168: F. P. Peterson, The Steenrod Algebra and its Applications: A conference to Celebrate N. E. Steenrod's Sixtieth Birthday VII, 317 pages. 1970. DM 22,–

Vol. 169: M. Raynaud, Anneaux Locaux Henséliens V, 129 pages. 1970. DM 16,–

Vol. 170: Lectures in Modern Analysis and Applications III. Edited by C. T. Taam. VI, 213 pages 1970. DM 18,–

Vol. 171: Set-Valued Mappings, Selections and Topological Properties of 2^X. Edited by W. M. Fleischman. X, 110 pages. 1970. DM 16,–

Vol. 172: Y.-T. Siu and G. Trautmann, Gap-Sheaves and Extension of Coherent Analytic Subsheaves. V, 172 pages. 1971. DM 16,–

Vol. 173: J. N. Mordeson and B. Vinograde, Structure of Arbitrary Purely Inseparable Extension Fields. IV, 138 pages. 1970. DM 16,–

Vol. 174: B. Iversen, Linear Determinants with Applications to the Picard Scheme of a Family of Algebraic Curves. VI, 69 pages. 1970. DM 16,–

Vol. 175: M. Brelot, On Topologies and Boundaries in Potential Theory. VI, 176 pages. 1971. DM 18,–

Vol. 176: H. Popp, Fundamentalgruppen algebraischer Mannigfaltigkeiten. IV, 154 Seiten. 1970. DM 16,–

Vol. 177: J. Lambek, Torsion Theories, Additive Semantics and Rings of Quotients. VI, 94 pages. 1971. DM 16,–

Vol. 178: Th. Bröcker und T. tom Dieck, Kobordismentheorie. XVI, 191 Seiten. 1970. DM 18,–

Vol. 179: Seminaire Bourbaki – vol. 1968/69. Exposés 347-363. IV, 295 pages 1971 DM 22,–.

Vol. 180: Séminaire Bourbaki – vol. 1969/70. Exposés 364-381. IV, 310 pages. 1971. DM 22,–

Vol. 181: F. DeMeyer and E. Ingraham, Separable Algebras over Commutative Rings. V, 157 pages. 1971. DM 16.–

Vol. 182: L. D. Baumert. Cyclic Difference Sets. VI, 166 pages. 1971 DM 16,–

Vol. 183: Analytic Theory of Differential Equations. Edited by P. F. Hsieh and A. W. J. Stoddart. VI, 225 pages. 1971. DM 20,–

Vol. 184: Symposium on Several Complex Variables, Park City, Utah, 1970 Edited by R. M. Brooks. V, 234 pages. 1971. DM 20,–

Vol. 185: Several Complex Variables II, Maryland 1970. Edited by J. Horváth. III, 287 pages. 1971. DM 24,–

Vol. 186: Recent Trends in Graph Theory. Edited by M. Capobianco/ J. B. Frechen/M. Krolik. VI, 219 pages. 1971. DM 18.–

Vol. 187: H. S. Shapiro, Topics in Approximation Theory. VIII, 275 pages. 1971. DM 22,–

Vol. 188: Symposium on Semantics of Algorithmic Languages. Edited by E. Engeler. VI, 372 pages. 1971. DM 26,–

Vol. 189: A. Weil, Dirichlet Series and Automorphic Forms. V, 164 pages. 1971. DM 16,–

Vol. 190: Martingales. A Report on a Meeting at Oberwolfach, May 17-23, 1970. Edited by H. Dinges. V, 75 pages. 1971. DM 16,–

Vol. 191: Séminaire de Probabilités V. Edited by P. A. Meyer. IV, 372 pages. 1971. DM 26,–

Vol. 192: Proceedings of Liverpool Singularities – Symposium I. Edited by C. T. C. Wall. V, 319 pages. 1971. DM 24,–

Vol. 193: Symposium on the Theory of Numerical Analysis. Edited by J. Ll. Morris. VI, 152 pages. 1971. DM 16,–

Vol. 194: M. Berger, P Gauduchon et E. Mazet. Le Spectre d'une Variété Riemannienne VII, 251 pages. 1971. DM 22,–

Vol. 195: Reports of the Midwest Category Seminar V. Edited by J.W. Gray and S. Mac Lane. III, 255 pages. 1971. DM 22,–

Vol. 196: H-spaces – Neuchâtel (Suisse)- Août 1970. Edited by F. Sigrist, V, 156 pages. 1971. DM 16,–

Vol. 197: Manifolds – Amsterdam 1970. Edited by N. H. Kuiper. V, 231 pages 1971 DM 20,–

Vol. 198: M. Hervé, Analytic and Plurisubharmonic Functions in Finite and Infinite Dimensional Spaces. VI, 90 pages. 1971. DM 16.–

Vol. 199: Ch. J. Mozzochi, On the Pointwise Convergence of Fourier Series. VII, 87 pages. 1971. DM 16,–

Vol. 200: U Neri, Singular Integrals. VII, 272 pages. 1971. DM 22,–

Vol. 201: J. H. van Lint, Coding Theory. VII, 136 pages. 1971. DM 16,–

Vol. 202: J. Benedetto, Harmonic Analysis on Totally Disconnected Sets. VIII, 261 pages. 1971 DM 22,–

Vol. 203: D. Knutson, Algebraic Spaces. VI, 261 pages. 1971. DM 22,–

Vol. 204: A. Zygmund, Intégrales Singulières. IV, 53 pages. 1971. DM 16,–

Vol. 205: Séminaire Pierre Lelong (Analyse) Année 1970. VI, 243 pages. 1971. DM 20,–

Vol. 206: Symposium on Differential Equations and Dynamical Systems. Edited by D. Chillingworth. XI, 173 pages. 1971. DM 16,–

Vol. 207: L. Bernstein, The Jacobi-Perron Algorithm – Its Theory and Application. IV, 161 pages. 1971. DM 16,–

Vol. 208: A. Grothendieck and J. P. Murre, The Tame Fundamental Group of a Formal Neighbourhood of a Divisor with Normal Crossings on a Scheme. VIII, 133 pages. 1971. DM 16,–

Vol. 209: Proceedings of Liverpool Singularities Symposium II. Edited by C. T. C. Wall. V, 280 pages. 1971. DM 22,–

Vol. 210: M. Eichler, Projective Varieties and Modular Forms. III, 118 pages. 1971. DM 16,–

Vol. 211: Théorie des Matroïdes. Edité par C. P. Bruter. III, 108 pages. 1971. DM 16,–

Vol. 212: B. Scarpellini, Proof Theory and Intuitionistic Systems. VII, 291 pages. 1971. DM 24,–

Vol. 213: H. Hogbe-Nlend, Théorie des Bornologies et Applications. V, 168 pages. 1971. DM 18,–

Vol. 214: M. Smorodinsky, Ergodic Theory, Entropy. V, 64 pages. 1971. DM 16,–